Cybernetic Shield

Cybernetic Shield: Securing the Future of Machine Intelligence explores the critical intersection between cybersecurity and the rapid advancements in artificial intelligence. This book addresses the growing need for robust security measures in AI-driven systems by presenting cutting-edge strategies and frameworks that ensure the integrity, trust, and resilience of machine intelligence. It delves into a wide array of emerging cybersecurity challenges—from securing AI-integrated learning platforms in digital education to safeguarding blockchain-based solutions and tackling vulnerabilities in IoT networks. The book offers practical insights into protecting against cyber threats that threaten the changing digital landscape, guaranteeing a future where intelligent technologies can be safely harnessed for innovation and advancement. It does this by drawing on both theoretical analysis and real-world case studies.

This book offers an in-depth exploration of the latest cybersecurity techniques tailored for AI applications. The chapters cover a range of topics, from securing machine learning platforms and mitigating privacy risks in AI-driven healthcare to implementing decentralized blockchain systems and safeguarding critical systems from AI-driven vulnerabilities. Readers will benefit from practical solutions to pressing cybersecurity challenges, expert insights on governance frameworks, and real-world examples of how emerging technologies like IoT, AI, and blockchain are reshaping the future of cybersecurity. The book serves as an essential resource for anyone looking to stay ahead of the curve in the rapidly evolving digital landscape.

Cybernetic Shield: Securing the Future of Machine Intelligence is specifically designed for professionals, researchers, and students in the fields of cybersecurity, AI, and digital technologies. It caters to individuals involved in securing AI systems, blockchain development, and digital infrastructures across various industries, including healthcare, finance, and education. The book is also highly relevant to policymakers, legal experts, and technology leaders who are seeking guidance on creating frameworks to protect the future of machine intelligence. With its comprehensive approach, it appeals to anyone interested in understanding the critical role of cybersecurity in the age of AI and machine learning.

Sustainable Industrial Engineering and Management Systems

Series Editors: Pushan Kumar Dutta, Chithirai Pon Selvan, Kewei Sha and Darshan Desai

This series provides a central hub to explore innovative industrial engineering and management tools, techniques, and frameworks that enable economic, social, and environmental sustainability across organizational systems and supply chains. With a rising global imperative for industries to dramatically reduce their resource usage and waste as well as create positive lifecycle impacts, the volumes in this series equip operations leaders to drive transformational change. Authored by world-renowned researchers and practitioners at the nexus of sustainability and operational excellence, the books will give an integrated, systemic perspective across both manufacturing and service ecosystems. Core technical topics include circular production, symbiotic supplier networks, product/process reengineering, IoT-enabled flexibility and control, safety advancements, and rigorous impact quantification methods amongst others. Application domains span smart factories, transportation fleets, healthcare networks, and built environment systems. Volumes feature actionable case studies from pioneers, high-relevance lessons for managers planning interventions, as well as skill-building exercises. By disseminating cutting-edge IEM knowledge, the books aim to cross-pollinate innovations that accelerate prosperous, responsible, and renewable value creation globally. While rigorously grounded in technical and scientific discourse, the series maintains an accessible style suited to technology strategists, engineers, business leaders, and applied policymakers.

Engineering the Circular Economy
Leveraging Fuzzy Logic for Implementation
Edited by Biswadip Basu Mallik, Gunjan Mukherjee, Rahul Kar, and Youqing Fan

Sustainable Healthcare Systems in Africa
Technologies, Practices, and Management
Edited by Wasswa Shafik, Adel Ben Youssef, Chithirai Pon Selvan and Pushan Kumar Dutta

Security Paradigms in 6G Smart Cities and IoT Ecosystems
Navigating the Future
Edited by Sushil Kumar Singh, Rajendrasinh B. Jadeja, Ashish Khanna, Pushan Kumar Dutta, and Bal Virdee

Cybernetic Shield
Securing the Future of Machine Intelligence
Edited by Arun Kumar Singh, Sandeep Saxena, and A.K. Haghi

For more information about this series, please visit: www.routledge.com/Sustainable-Industrial-Engineering-and-Management-Systems/book-series/SIEMS

Cybernetic Shield

Securing the Future of Machine Intelligence

Edited by
Arun Kumar Singh, Sandeep Saxena,
and A.K. Haghi

CRC Press
Taylor & Francis Group
Boca Raton London New York

CRC Press is an imprint of the
Taylor & Francis Group, an **informa** business

First edition published 2026
by CRC Press
2385 NW Executive Center Drive, Suite 320, Boca Raton FL 33431

and by CRC Press
4 Park Square, Milton Park, Abingdon, Oxon, OX14 4RN

CRC Press is an imprint of Taylor & Francis Group, LLC

ISBN: 978-1-032-93643-7 (hbk)
ISBN: 978-1-032-93772-4 (pbk)
ISBN: 978-1-003-56758-5 (ebk)

DOI: 10.1201/9781003567585

Typeset in Times
by Deanta Global Publishing Services, Chennai, India

Contents

Preface

The last decade has witnessed an unprecedented convergence of artificial intelligence, cybersecurity, and pervasive sensing, a fusion that is transforming critical infrastructure as thoroughly as it is reframing everyday life. Algorithms that once lived inside controlled laboratories now authenticate legal signatures, adjudicate loan approvals, and protect marine habitats, while the data that fuels them flows continuously from edge devices embedded in homes, hospitals, and city streets. *Cybernetic Shield: Securing the Future of Machine Intelligence* is therefore more than a title; it is the central thesis of this volume. Our contributors write at a moment when large language models adjudicate contracts, federated algorithms diagnose disease, and edge devices orchestrate entire factories. Yet these same systems, if left unguarded, can amplify bias, leak private data, or collapse under targeted attack. The collection you hold was conceived to meet that dual reality head-on, offering an integrated playbook that binds theoretical rigor, engineering practice, and governance foresight into a single narrative.

The volume *Cybernetic Shield: Securing the Future of Machine Intelligence* is organized into three distinct sections, each focusing on different facets of AI, cybersecurity, and pervasive sensing. Section I: Foundational Perspectives in AI and Cybersecurity (Chapters 1–7) lays the groundwork by exploring core concepts like AI governance, cybersecurity frameworks, and emerging challenges such as signature forgery detection and AI-integrated learning platform security. Moving from theory to practical implementation, Section II: AI-Based Surveillance, Intrusion Detection, and Privacy (Chapters 8–14) delves into real-world applications, including election integrity, federated learning in healthcare, and machine learning models for detecting cyber vulnerabilities. Finally, Section III: Intelligent Systems Using IoT and Edge AI (Chapters 15–21) extends the conversation into cutting-edge areas such as blockchain, cloud-based applications, IoT-based security systems, and AI-driven underwater detection, offering insights into how intelligent systems can be integrated into the physical world to create resilient and secure environments. This progression from foundational principles to practical, field-tested prototypes provides a comprehensive framework for understanding and addressing the evolving cybersecurity challenges posed by modern AI systems.

Individual chapters interlock like plates in a composite armor. A legal framework for safeguarding machine intelligence feeds directly into an empirical study on securing AI-integrated learning platforms; research on privacy-preserving transfer learning for brain tumor classification converses with a later examination of open-access neuro-datasets, together illustrating how transparency and confidentiality can coexist. Blockchain-driven audits of big-tech vulnerabilities prepare the ground for a cloud-scaling blueprint that optimizes resource allocation without sacrificing trust. Whether analyzing YOLO-Tiny optimizers for airport gate assignments or unveiling metasurface antennas for wearable healthcare, each contribution answers a facet of the volume's central mandate: intelligence fortified by design, not patched after deployment.

The individual chapters form an intentional tapestry rather than a simple anthology. A discussion of legal guardrails for machine intelligence naturally precedes an empirical study on securing AI-integrated learning platforms, allowing readers to trace how policy imperatives translate into architectural choices. Work on privacy-preserving transfer learning for brain tumor classification converses with a later contribution on open-access brain-computer interface datasets, collectively demonstrating how transparency can coexist with confidentiality. Similarly, a chapter that dissects election-integrity threats foreshadows a blockchain-based framework for supply-chain transparency, revealing common cryptographic primitives across civic and commercial domains. Whether introducing a YOLO-Tiny-based gate-assignment optimizer for crowded airports or unveiling a metasurface antenna that powers wearable health sensors, each study advances a shared goal: to engineer intelligence without surrendering trust. Because threats and opportunities cross organizational boundaries, the *Cybernetic Shield* is addressed to a layered readership. This volume speaks to multiple communities. Researchers will find fresh theoretical models, benchmark datasets, and reproducible code samples that invite further experimentation. Security professionals and system architects can adapt the intrusion-detection blueprints, edge-AI deployment checklists, and cloud-scaling heuristics to their own production environments. Policymakers will discover analyses that translate abstract governance principles into enforceable standards, while educators can draw on case studies—such as dropout-prediction analytics and NLP-driven cyberbullying detection—to illustrate real-world impact in the classroom. We assume only a foundational familiarity with computing concepts; the chapters deliberately scaffold complexity so that readers from diverse backgrounds can enter at their own level and still converge on a holistic understanding of the field.

Gratitude is also due to the production team, who tamed intricate diagrams and code listings, and to our families, whose patience transformed late-night revisions into a labor of shared mission. We invite you to test, challenge, and extend the ideas collected here, and we hope the *Cybernetic Shield* becomes both a reference and a provocation—evidence that bold innovation and meticulous stewardship can, and must, evolve in unison.

Editors

Arun Kumar Singh is professor and Deputy Head of the Department of Computer Science and Engineering at Greater Noida Institute of Technology (GNIOT), India. He has completed specialized training in Intellectual Property and the Patent Cooperation Treaty (PCT) through the World Intellectual Property Organization (WIPO) Academy, enhancing his understanding of legal aspects related to innovation protection. Dr Singh has a diverse professional background, having worked in academia, research, and industry. In addition to his roles at GNIOT, he worked at the University of Allahabad and MNNIT. His previous positions include professor, associate professor, and head of department at esteemed institutions such as GNIOT, GL Bajaj Institute of Technology and Management, and the Institute of Engineering and Rural Technology.

Dr Singh has made significant contributions to research, with publications in various SCI journals, international conferences indexed in Scopus, and Scopus book chapters. He has worked on projects funded by organizations such as CSIR and ISEA. He is also the author of textbooks, patent publications, and several grants. His expertise spans cybersecurity, cryptography, machine learning, and computer forensics. Additionally, he is AWS certified and an expert in National Board of Accreditation (NBA) and National Assessment and Accreditation Council (NAAC) accreditation processes, with a focus on outcome-based education.

Sandeep Saxena is a professor at the School of Computer Science and Engineering, IILM University, Greater Noida, India. He earned a PhD in computer science and engineering from NIT Durgapur, West Bengal, India. With over 16 years of teaching and research experience, he specializes in blockchain technology and cloud computing, particularly focusing on security and privacy. His work includes research in cloud computing architecture design and access control techniques.

He has actively contributed to over 20 international conferences, serving as a keynote speaker, the organizing secretary, the organizing chair, and the session chair. He has authored four technical books and published more than 50 research papers in top international journals and conferences. His work is indexed by renowned platforms such as Scopus, SCIE, Google Scholar, and DBLP, with high-impact publications.

He holds three patents, with more than ten additional ones published. As a senior member of IEEE and IAASSE and a lifetime member of CSI and CRSI, he is actively involved in professional societies. His roles extend to research and development, university syllabus design, and guiding student innovation and incubation projects. Additionally, he has served as a reviewer for leading journals and conferences, including those published by IEEE, Springer, and other Scopus-indexed journals, maintaining the integrity of academic work. He has also contributed to accreditation bodies such as NAAC and NBA, helping improve the educational landscape.

A.K. Haghi is a retired professor and has written, co-written, edited, or co-edited more than 1000 publications, including books, book chapters, and papers in refereed journals. Professor Haghi holds a BSc in urban and environmental engineering from the University of North Carolina (USA) and holds two MSc degrees, one in mechanical engineering from North Carolina State University (USA) and another in applied mechanics, acoustics, and materials from the Université de Technologie de Compiègne (France). He was awarded a PhD in engineering sciences by the Université de Franche-Comté (France). Professor Haghi's extensive educational background and supervisory roles underscore his expertise and contributions to the field of engineering sciences. He is an appointed honorary research associate (HRA) at the University of Coimbra, Portugal, and a regular reviewer of leading international journals.

Contributors

M. Aanandalakshmi
Velammal Engineering College
Chennai, India

Malikov Ziyodullo Abdurayim
Samarkand State University
Samarkand, Uzbekistan

David Akpuluma
Tomsk Polytechnic University
Tomsk, Russia

Christo Ananth
Samarkand State University
Samarkand, Uzbekistan

Qodirov Asliddin Asomiddin
Samarkand State University
Samarkand, Uzbekistan

G. Balamurugan
SRM Institute of Science and Technology, Kattankulathur Campus
Chengalpattu, India

Niha Kamal Basha
Vellore Institute of Technology
Vellore, India

Vikash Deep Bhaskar
Madan Mohan Malaviya University of Technology
Gorakhpur
Uttar Pradesh, India

Shashi Bhushan
Universiti Teknologi PETRONAS
Seri Iskandar, Malaysia

Prasanthi Boyapati
SRM University
India

Vanisha Buragohain
Amity University Kolkata
Kolkata, India

Rakhmanov Buribay
Samarkand State University
Samarkand, Uzbekistan

Kalyan Chatterjee
Amity University Kolkata
Kolkata, India

Rajesh Dey
IIUM Institute of Islamic Banking and Finance
Selangor, Malaysia

Rahaf Mohammedovna Douhan
National Research Tomsk State University
Tomsk, Russia

Neda Firoz
Tomsk State University
Tomsk, Russia

Swetha Ghanta
SRM University
Andhra Pradesh
India

G. S. Pradeep Ghantasala
Alliance College of Engineering and Design
Alliance University
Bengaluru, India

Priyanka Gorai
SRM Institute of Science and Technology, Kattankulathur Campus
Chengalpattu, India

Ankit Gupta
SRM Institute of Science and Technology
Kattankulathur, India

Lochini J.
Sri Sairam Engineering College
Chennai, India

Pratibha Vijay Jadhav
MIT-ART, Design and Technology
Pune, India

Kavita Kanabar
Vivekanand Education Society's Institute of Management Studies & Research
Mumbai, India

Eirtty Kapoor
Delhi Technical Campus
Greater Noida
Uttar Pradesh, India

Mudassir Khan
Tanumah, King Khalid University
Abha, Saudi Arabia

S. Kolangiammal
SRM Institute of Science and Technology
Kattankulathur
Tamil Nadu, India

Dhananjay Kumar
Bennett University
Greater Noida
Uttar Pradesh, India

Babli Kumari
Government Polytechnic Latehar
Latehar, India

Rajesh Kumhar
Mizoram University, Aizawl
Mizoram, India

Rupali Mahajan
Vishwakarma Institute of Technology
Pune, Maharashtra
India

Tarkesh Kumar Mahato
Government Polytechnic Ranchi
Ranchi, India

Nazarov Fayzullo Makhmadiyarovich
Samarkand State University
Samarkand, Uzbekistan

Ziyodullo Abdurayim Malikov
Samarkand State University
Samarkand, Uzbekistan

Navonita Mallick
School of Law, KIIT-DU
Bhubaneswar, India

M. Maneesh
SRM Institute of Science and Technology.
Kattankulathur
Tamil Nadu, India

Divya Mishra
Infrastructure University Kuala Lumpur (IUKL)
Kuala Lumpur (IUKL), Malaysia

R. Neharika
Velammal Engineering College
Chennai, India

Kalaiselvi P.
Sri Sairam Engineering College
Chennai, India

Vijay Krishna Pallaw
G.L. Bajaj College of Technology & Management
Greater Noida
Uttar Pradesh, India

M. Parvathy
Velammal Engineering College
Chennai, India

Subhra Prosun Paul
SRM University
India

Shaik Karimullah
Annamacharya University
Rajampet, Kadapa
Andhra Pradesh, India

Ashok Kumar Pradhan
SRM University
India

Rajesh Prasad
Stella Mary's College of Engineering
Azhikal PO
Kanyakumari, India

Ritwik Priya
Shri Jagdishprasad Jhabarmal Tibrewala University
Jhunjhunu, India

Yusupov Ozod Rabbimovich
Samarkand State University
Samarkand, Uzbekistan

Laavanya Rachakonda
University of North Carolina Wilmington
North Carolina, USA

Subhajit Roy
Cybersecurity & DF, ISOEH, Kolkata & Dept. of EE NIT
Silchar, India

Subhasish Roy
Amity University Kolkata
Kolkata, India

Akhatov Akmal Rustamovich
Samarkand State University
Samarkand, Uzbekistan

Yoga Lakshmi S.
Sri Sairam Engineering College
Chennai, India

Bhawna Sachdeva
Greater Noida Institute of Technology
Greater Noida
Uttar Pradesh, India

Shashwata Sahu
School of Law, KIIT-DU
Bhubaneswar, India

Debabrata Samanta
Rochester Institute of Technology—RIT Global Campus
Kosovo, Europe

Sushil Kumar Saroj
Madan Mohan Malaviya University of Technology
Gorakhpur
Uttar Pradesh, India

Nida Shakeel
Madan Mohan Malaviya University of Technology
Gorakhpur
Uttar Pradesh, India

Shagufta Shakeel
Madan Mohan Malaviya University of Technology
Gorakhpur
Uttar Pradesh, India

Sharmila
Raj Kumar Goel Institute of Technology
Ghaziabad
Uttar Pradesh, India

R. Shreya
SRM Institute of Science and Technology
Kattankulathur
Tamil Nadu, India

Rishabh Dev Shukla
Chandigarh University
Mohali, India

Arjun Singh
Greater Noida Institute of Technology
Greater Noida
Uttar Pradesh, India

Anuj Singh
SRM Institute of Science and Technology, Kattankulathur Campus
Chengalpattu, India

Sandeep Kumar Singh
G.L. Bajaj College of Technology & Management
Greater Noida
Uttar Pradesh, India

Sonam Sirohi
Greater Noida Institute of Technology
Greater Noida
Uttar Pradesh, India

V. Subbulakshmi
Velammal Engineering College
Chennai, India

Abduraxmonov Muhammad Sulaymon
Samarkand State University
Samarkand, Uzbekistan

Aishwarya T.
Sri Sairam Engineering College
Chennai, India

Mohammad Sadman Tahsin
Texas A&M University–Kingsville
Kingsville
Texas, USA

Rohit Kumar Tiwari
Madan Mohan Malaviya University of Technology
Gorakhpur
Uttar Pradesh, India

Priyanka Tyagi
Raj Kumar Goel Institute of Technology
Ghaziabad
Uttar Pradesh, India

Arun V.
Sri Sairam Engineering College
Chennai, India

1 Cyber Defense Strategies for Ensuring Electoral Trust and Integrity

Neda Firoz, Arun Kumar Singh, and David Akpuluma

1.1 INTRODUCTION

Elections are fundamental to democratic governance and are crucial in determining a nation's direction. Nevertheless, the dominant ballot-paper voting systems persistently encounter trust challenges across substantial demographics. Consequently, e-voting has surfaced as an attractive option, with several nations choosing to adopt it worldwide [1]. Although e-voting systems have numerous benefits, they also include specific obstacles. A trivial vulnerability can result in significant alterations to vote outcomes. Recent years have seen attempts to transform the e-voting paradigm by utilizing emerging technologies such as biometrics and blockchain [2]. This chapter offers an internet-based voting system that utilizes blockchain technology and biometric identification methods. We employ biometric modalities, including fingerprint and facial recognition, for voter authentication, utilizing the Hyperledger Fabric platform as the blockchain network to provide a secure, transparent, and tamper-evident voting record [3]. We illustrate the suggested method with 100 participants in a controlled environment where we gather biometric data. The findings reveal that 87% of participants successfully registered using biometrics, while 88% voted utilizing either a combination of voter ID and fingerprint or voter ID and facial recognition. Our findings indicate that the suggested system enables voters to access it effortlessly and automates identity verification processes while maintaining a safe, decentralized, and distributed database network that upholds transparency. A subsequent study will be conducted in partnership with election officials and voters to enhance the system in practical contexts.

1.1.1 Voter Confidence as the Cornerstone of Digital Elections

In any electoral process, the trust of voters is paramount; the legitimacy of both the voting system and the elected body hinges on this foundational confidence. As the adoption of cloud-based voting platforms accelerates, transparency has [4] emerged as a non-negotiable criterion for ensuring voter trust and participation. If voters

DOI: 10.1201/9781003567585-1

perceive a system to be insecure or opaque, their likelihood of engagement drops significantly. To address this, modern digital voting platforms must incorporate mechanisms such as verifiable voter receipts, which confirm that a vote has been both cast and recorded correctly [5]. Moreover, transparent audit methods allow both voters and administrators to verify the results on their own and can help promote a culture of accountability. Unique voter IDs are an important tool for preserving electoral integrity, since [6] they help to ensure that each vote belongs to just one legitimate person by preventing digital ballot stuffing. Any discrepancies that emerge will be instantly detected by detailed audit logs, which make it possible to respond to the incidents in real time and perform forensic analysis [7]. The continued acceleration of platforms such as ElectionBuddy, especially within regions such as Asia and other emerging areas, demonstrates the emerging global acceptance of secure and transparent online voting options. This trend highlights the transformative power of electronic voting systems to increase voter turnout and strengthen democratic institutions around the globe.

1.1.2 Real-Time Threat Detection and Incident Response

In the fast-changing cybersecurity environment, continuous monitoring of threats and instant response in case of incidents became important to ensure the availability and integrity of digital infrastructures. The increasing complexity and frequency of cyberattacks cannot be effectively countered with traditional endpoint security solutions based on analysis after the event or scheduled scans. Real-time threat detection is a process that keeps an eye on systems, networks, and user activities to detect any incidents or potential threats as they transpire [8]. This proactive approach taps into advanced technologies, including AI, ML, and big data analytics, to process large amounts of security data in real time and provide early warning signs of malicious activities such as malware insertion, unauthorized entry, denial-of-service hacks, and insider abuses.

By contrast, incident response is the methodical process by which a company or organization addresses and mitigates the fallout from a cyberattack once that attack has been detected. An effective incident response program mitigates the extent of damage and recovery efforts, but it also assists in preserving digital evidence, regulatory compliance, and resuming operations quickly. Security information and event management (SIEM), intrusion detection and prevention systems (IDPS), and endpoint detection and response (EDR) platforms are some of the technologies in which real-time monitoring with automatic response is conducted. The addition of real-time threat feeds, behavior analytics, and automated playbooks greatly extends an organization's capability to identify and eliminate threats in real time. In addition, the rise of cloud-centric and distributed infrastructures has also resulted in the need for scalability and intelligence. Furthermore, as organizations shift towards cloud-based and distributed environments, the importance of scalable and intelligent real-time defense mechanisms has grown exponentially. Ultimately, investing in real-time detection and rapid response strategies is not just a security imperative—it is foundational to safeguarding data assets, maintaining customer trust, and ensuring business continuity in today's hyperconnected world.

1.2 LITERATURE SURVEY

Blockchain technology has been suggested as a remedy for the security and transparency issues inherent in e-voting systems [9]. Numerous studies have examined the incorporation of blockchain technology into electronic voting systems, yielding encouraging outcomes. Researchers suggested an integrated e-voting system that effectively utilized blockchain technology inside a supervised non-remote internet voting framework, providing end-to-end verifiability. The system implemented the Auditable Blockchain Voting System (ABVS), and later assessments demonstrated that the ABVS surpassed previous electronic voting systems regarding security and reliability. Additional research has employed diverse blockchain frameworks, including Hyperledger Fabric and Ethereum, to create safe e-voting systems. These research studies have shown that blockchain technology can significantly enhance transparency and security in public applications.

Researchers have proposed voting frameworks that employ blockchain algorithms developed through cryptography approaches. They have illustrated the efficacy and usefulness of e-voting systems and the possibility for substantial enhancement of electronic voting via the implementation of blockchain [10]. A recent study performed a comprehensive analysis of traditional and blockchain-based voting systems, assessing the current landscape of blockchain-focused voting research. The analysis indicated that blockchain technology could potentially mitigate numerous challenges that presently obstruct openness in our electoral system. The researchers determined that for a functional blockchain-based electronic voting system, it is essential to guarantee the security of remote participation and to enhance transaction speed for scalability. The current frameworks [11] necessitate substantial improvements for effective incorporation into voting systems.

1.2.1 Research Significance and Gap Analysis

Current e-voting systems have advanced considerably in incorporating upcoming technologies such as blockchain. Nonetheless, these systems have insufficiently resolved voter identity and verification concerns, along with potential vulnerabilities that may permit substantial alterations in voting results. The current research also did not concentrate on web-based internet voting. This vacuum in the literature signifies a vital area where our research aims to provide a substantial contribution.

1.2.2 Regulatory Adherence and Examination in Digital Elections

In democratic nations, adherence to regulations and strong auditing systems are essential for preserving the integrity, transparency, and public confidence in electoral processes. Every nation—and frequently each jurisdiction within a nation—implements specific legislation to oversee elections. In the United States [13], individual states adhere to distinct election regulations, resulting in a complicated compliance landscape for election authorities. To traverse this domain, solutions such as Privileged Access Management (PAM) are being utilized to fulfill the rigorous cybersecurity standards linked to elections. PAM solutions provide secure management of access to essential systems and support thorough auditing and monitoring of privileged user actions. This is crucial for complying with the requirements established by the U.S. Election Assistance Commission (EAC), which requires the adoption of security measures that ensure the integrity of voting systems. Solutions such as senhasegura enable election authorities to monitor, document, and audit all actions performed by users with higher access, thereby ensuring complete transparency and accountability. This detailed oversight guarantees compliance with regulatory requirements and is crucial in a time when election results face heightened public and political examination [12]. As digital election technologies proliferate, such auditing capabilities are essential for preserving democratic legitimacy and maintaining the rule of law in electoral governance [13]. Table 1.1 shows the comparative study of electronic voting and paper ballots.

1.3 MITIGATING INSIDER THREATS IN ELECTION SECURITY

Safeguarding election systems in contradiction of insider threats is similarly critical to defending them from external attacks. Without appropriate oversight, employees or contractors with elevated privileges may exploit their access for malicious purposes. PAM mitigates risks by categorizing users into groups according to their access requirements, ensuring that each individual possesses only the permissions necessary for their role. PAM systematically monitors and logs each action performed, facilitating real-time analysis to identify any anomalous behavior. By monitoring and analyzing user activities, election officials can enhance the safety of the process by recognizing potential insider threats before they escalate. This measure is designed to maintain the integrity of the election process.

1.3.1 Multifactor Authentication: The First Line of Defense

Election security demands utmost diligence, especially through the principle of least privilege. It is a key strategy for preventing unwanted penetration when configuring services.

It ensures that the users have only the access required to perform their function. PAM helps to keep this policy in place by constantly observing user activity and guarding sensitive data. Also, assuming potential attacks from inside and outside of the organization, the Zero Trust concept design [14] can increase the organization's security posture. To offset threats and protect election systems from potential threats, PAM allows election authorities to operate Zero Trust through verifying every request for access and continuously monitoring the behavior of users.

TABLE 1.1
A Comparative Study of Electronic Voting and Paper Ballots

Parameter	Electronic Voting (E-Voting)	Paper Ballot Voting
Speed and efficiency	Faster processing of votes, immediate tallying.	Slower processing due to manual counting of votes.
Cost	High initial setup costs (infrastructure, technology, maintenance).	Lower initial costs, but may require ongoing printing costs.
Security	Potential vulnerabilities in software, hardware, and networks. Can be enhanced with encryption, biometrics, and secure protocols.	Vulnerable to human errors, manipulation, or fraud (e.g., vote tampering, ballot box stuffing).
Accessibility	More accessible for disabled individuals (e.g., touchscreens, voice commands).	May be challenging for people with disabilities (e.g., physical limitations).
Transparency	Can be more difficult to audit or verify, especially if systems are proprietary or closed-source.	Transparent, as paper ballots are physically counted and can be manually verified.
Voter privacy	Advanced encryption can protect voter privacy, but concerns about data breaches and hacking exist.	Voter privacy is relatively secure, though it may be compromised in certain situations (e.g., public voting stations).
Ease of use	User-friendly interfaces, but may require training for new users.	Simple and intuitive for voters, familiar to most people.
Reliability	Technical issues (e.g., system malfunctions, power outages) can disrupt the process.	Less prone to technical failure, but physical damage (e.g., destroyed ballots, weather conditions) can pose a problem.
Scalability	Easily scalable for large populations or remote voting.	More challenging to scale in large elections; requires a lot of manpower and logistics.
Environmental impact	Requires electronics and power, leading to e-waste and energy consumption.	Less environmental impact but uses paper, contributing to deforestation and waste.
Fraud and manipulation risks	Potential for cyberattacks, hacking, or system manipulation. Requires robust cybersecurity measures.	Higher risk of human errors, ballot box tampering, and vote miscounts.
Auditability	Difficult to manually audit without transparency mechanisms (e.g., paper trails).	Easy to audit due to the physical presence of ballots, with clear records.
Public trust	Varies by country; can be affected by technical issues, transparency, and security concerns.	Generally higher public trust, but prone to human error and fraud.
Long-term sustainability	Long-term costs related to system upgrades, cybersecurity, and maintenance.	Less expensive in the long run but could become inefficient for large-scale elections.

1.3.2 Least Privilege and Zero Trust

Election security is quite important. One is that of least privilege—the focus that you can track down unauthorized access. It also means users only get the access they absolutely require to do their work. One of these rules is enforced by the Pluggable Authentication Module, which is intended to watch everything that users do so that no sensitive information is put at risk. Moreover, according to [15], the Zero Trust model contributes to cybersecurity solutions, since it builds on the assumption that threats might exist both externally and internally. Election officials must now assume a Zero Trust stance with PAM. In a Zero Trust environment, you don't trust any request to access, and you observe what they are doing all the time to help mitigate risk and protect election systems from any potential threat.

1.4 CONCLUSION

The primary objective in designing e-voting is to focus on the characteristics and drawbacks of paper-based electronic voting. Electronic voting is a convenient system with many benefits, such as high speed and efficiency, and it has the potential for large-scale voting systems. Additionally, it is the voting system for the disabled community. But it also raises issues regarding security, transparency, and the potential for technical errors or cyberattacks that could undermine public trust. Another alternative that is coming into focus is the paper ballot, which in some configurations provides a more rugged and trustworthy means of counting votes and is often considered easier to understand and less prone to manipulation during audits. But such an approach also carries practical downsides, such as a prolonged process for tallying the votes, increased costs and the potential for fraud and human error. "With bottles, people have seen and touched them, so the public can feel comfortable with them as packaging." In the end, the decision for e-voting or paper-based voting is based on the particular requirement of the election, existing infrastructure, security concerns, and the objective of sustaining public confidence. They believe a mix of the best of both systems—such as a paper-based voting with an electronic system for a sort of "post election" audits—would offer a nice way of achieving both security and efficiency.

REFERENCES

1. Alam, M., Yusuf, M. O., & Sani, N. A. (2020). Blockchain technology for electoral process in Africa: a short review. *International Journal of Information Technology, 1*(3), 861–867. https://doi.org/10.1007/s41870-020-00440-w
2. Alvarez, R. M., Levin, I., Pomares, J., & Leiras, M. (2013). Voting made safe and easy: the impact of e-voting on citizen perceptions. *Political Science Research and Methods, 1*(1), 117–137. https://doi.org/10.1017/psrm.2013.2
3. Barber, B. R. (2001). Which technology for which democracy? Which democracy for which technology?. *International Journal of Communications Law and Policy, 6*(8), 7.
4. Boucher, P. (2016). What if blockchain technology revolutionised voting? In *EPRS (European Parliamentary Research Service)*. Retrieved December 6, 2024.

5. Chafiq, T., Azmi, R., & Mohammed, O. (2024). Blockchain-based electronic voting systems: a case study in Morocco. *International Journal of Intelligent Networks*, *5*, 38–48.
6. Chen, L., Cong, L. W., & Xiao, Y. (2020). A brief introduction to blockchain economics. In *World Scientific eBooks* (pp. 1–40).
7. Hajian Berenjestanaki, M., Barzegar, H.R., El Ioini, N., & Pahl, C. (2023). Blockchain-based e-voting systems: a technology review. *Electronics*, *13*(1), 17.
8. Haider, Z., & Daniel, M. (2024). Blockchain technology for enhanced security and trust in e-voting systems.
9. Olaniyi, O.O. (2024). Ballots and padlocks: Building digital trust and security in democracy through information governance strategies and blockchain technologies. Available at SSRN 4759942.
10. Shat, M.Z. (2025). Public perception of voting systems: evaluating trust in traditional vs. e-voting. *International Journal for Research in Applied Science and Engineering Technology*, *13*(1), 299–305.
11. Ohize, H.O., Onumanyi, A.J., Umar, B.U., Ajao, L.A., Isah, R.O., Dogo, E.M., Nuhu, B.K., Olaniyi, O.M., Ambafi, J.G., Sheidu, V.B., & Ibrahim, M.M. (2025). Blockchain for securing electronic voting systems: a survey of architectures, trends, solutions, and challenges. *Cluster Computing*, *28*(2), 132.
12. Singh, R.K., Chaudhary, L., Kumar, R., Kumar, A., Atul, A., & Bhardwaj, E.S. (2024, November). Securing elections with blockchain: a review of blockchain based e-voting systems. In *2024 International Conference on Advances in Computing, Communication and Materials (ICACCM)* (pp. 1–6). IEEE.
13. Pandya, S. (2022). advanced blockchain-based framework for enhancing security, transparency, and integrity in decentralised voting system. *International Journal of Advanced Research in Science, Communication and Technology*, 2(1), 865–876.
14. Raza, S.S., Khan, M.A., Shaikh, M.R., Alam, S., Talha, M., & Razzaq, E. (2025). A comparative study of blockchain-based e-voting systems: challenges, techniques, and implementation. *Spectrum of Engineering Sciences*, *3*(5), 356–372.
15. Ainur, J., Gulzhan, M., Amandos, T., Venera, R., Bulat, S., Zauresh, Y., & Aizhan, S. (2024). The impact of blockchain and artificial intelligence technologies in network security for e-voting. *International Journal of Electrical & Computer Engineering (2088–8708)*, *14*(6), 6723–6733.

2 Cyber Barriers
Leveraging IoT-Driven Security Systems for Enhanced Wildlife Protection

Eirtty Kapoor, Arun Kumar Singh, Neda Firoz, and Rahaf Mohammedovna Douhan

2.1 INTRODUCTION

Dream of a world where wild animals roamed freely and luxuriant foliage grew in the metropolis. This may sound like a utopian fantasy, but a security fence can make it happen. You read correctly! Security fences, commonly used to keep humans out, may help preserve wildlife and plants [1]. The use of security fences in the environment and their significance in re-establishing the harmony between people and the natural world are the focus of this book chapter. Initially, we shall examine security fences, while considering protective fence for wildlife preservation. Individuals and animals are obstructed by a barrier for safety purposes. Building a secure enclosure on your property helps protect local animals from predators, vehicular traffic, and other urban hazards. In regions abundant with deer, a barrier fence may prevent their intrusion into roadways and residential areas, therefore reducing accidents and injuries. Fencing may also facilitate the construction of animal pathways for secure movement. A security fence may also assist in protecting endangered animals. Establishing a safe gap between delicate ecosystems and human activity helps save rare and endangered species. Fencing can save endangered animals, such as the Sumatran tiger, from being killed during building activities [2]. This is the method by which various species are preserved, ensuring the protection of the area's biodiversity. It may also facilitate natural plant growth by serving as a security barrier. Fencing may deter invasive species from encroaching into local ecosystems by delineating the boundaries between natural habitats and human-occupied regions. Invasive flora and fauna may vie for resources with indigenous species, detrimental to ecosystems and biodiversity. Refuse or pets that escape are prevalent means by which invasive species may infiltrate your yard, and a security fence may mitigate this risk. Fencing also facilitates the natural growth of plants. It may prevent invasive species from accessing local ecosystems by isolating them from human habitats and workplaces. Invasive flora and fauna may vie with indigenous species

DOI: 10.1201/9781003567585-2

for resources, potentially detrimental to ecosystems and biodiversity. They may infiltrate your garden via discarded garbage and escaping pets. However, the installation of a security fence may mitigate such concerns. A security fence may improve deteriorated regions. Enclosing an abandoned industrial site with a fence allows for the safe recolonization of native flora and wildlife. This can regenerate the local ecology, increasing biodiversity and wildlife habitat. Last, security fences have benefits beyond the environment. Installing a safe fence around your home gives you peace and tranquility away from the city. The fence's secure sanctuary may surprise you with the animals you might see in your backyard.

2.2 LITERATURE SURVEY

Animals are destroying crops and agricultural areas globally. Various devices have been deployed to reduce animal incursion. The integration of AI with IoT has significantly contributed to addressing this issue. This section discusses endangered and hazardous animal detection and preventive studies and technology [3]. This fencing method is cheaper than standard fencing, barbed wires, or timber barriers and requires less installation. The cost of an electric fence system varies based on design and geography.

In farming, fences keep wild animals away from crops, which saves money and prevents fights between people and animals. It prevents animals from harming agriculture. Innovative fence systems, such as sensor-based and AI-driven ones, improve farm security and encourage cooperation while being environmentally sustainable [4]. Here we examine the development of an IoT-based security device to safeguard agricultural regions from wild animals. The device uses laser and vibration sensors with microcontrollers to detect field disruptions and unauthorized access. If trespassing or intrusion is detected, the system sends notifications via SMS, visual indications, or sound. Hardware configuration contains Arduino microcontrollers, conversation modules, and signal-giving devices, combined through circuit designs and tested for dependability. The tested technology effectively detects and alerts farm owners about hazards, ensuring trustworthy farm safety solutions. Researchers discussed agricultural loss by wild animals and the limitations of traditional methods such as scarecrows and fences. The chapter discusses current options, including IoT-based buildings, ultrasonic sensors, and AI-driven technologies, using Arduino, PIR sensors, and cameras to detect and prevent animal invasions. This innovative device aims to protect plants, boost harvests, and support farmers financially through cost-effective and sustainable methods. According to the study, a non-deadly, cost-effective, and adaptive system using PIR sensors, ultrasonic sensors, GSM modules, and sound recording modules can reduce wild animal intrusions in agricultural fields without harming the environment [5]. This gadget provides 360-degree entry detection for animals and birds using spinning ultrasonic sensors, real-time tracking with GSM indications, and sound playbacks to deter animals without injury. Combining Arduino IDE with ATMEGA 328P enables innovative automation, less human supervision, and enhanced performance, overcoming past system limitations, including limited detecting range, animal injury, and lack of adaptability [7].

Geofencing, a virtual boundary for a geographical region, is researched for tracking and automating GPS usage [8]. Previous investigations and geofencing strategies, including proximity to elements of interest, route adherence, and geofenced locations for cell item tracking, are reviewed. The primary goal is to create strong buildings that prevent trespassing and protect property and crops from damage. It optimizes Android geofencing utilizing location APIs for efficiency and accuracy. Developed an automated system to protect crops from wild animals, solving human-animal conflicts. Ultrasonic sensors and thermal cameras with deep learning algorithms help find and identify animals. High-frequency noises scare animals, while IoT integration informs farmers and authorities. Solar electricity and batteries power the gadget sustainably, ensuring cost-effective and efficient operation. The goal of this innovation is to minimize agricultural damage, protect farmers' livelihoods, and foster harmony between people and wildlife. The importance of effective surveillance and safety measures in agricultural areas to prevent crop damage from animal invasions is emphasized. Costly, risky, and unsuccessful methods include physical barriers, chemical repellents, and manual surveillance.

Table 2.1 provides a concise comparison of each technique, showing both its benefits and the limitations that can make it less effective or impractical for certain farming operations.

The research used IoT devices, deep learning, and M2M communication to create an intruder-detection system. The authors developed a modified CNN architecture for animal recognition. The IoT application used the M2M protocol to send a message to farm owners about animal movements. The team used image processing and ML algorithms to identify animal encroachment in agricultural fields. They used SVMs to classify pictures using Gabor features taken from the images.

2.3 CHALLENGES

There have been a number of obstacles to the implementation of IoT-integrated fence for animal security during the last five years, including both operational and technological ones. One notable problem has been the expensive initial installation costs, which remain a barrier, particularly in large-scale or isolated conservation areas with little financing. The fact that Internet of Things (IoT) gadgets depend so heavily on reliable, long-term power has also been a cause for concern in faraway places. Interruptions in data transmission caused by network connectivity in rural or forested areas have also led to lapses in real-time monitoring [11]. System failures have been induced by environmental factors, such as severe weather and interactions with wildlife. Data privacy and critical system protection have become major concerns due to the increased reliance on the IoT and the corresponding rise in the likelihood of hacking and cybersecurity breaches. These state-of-the-art technologies have proven to be difficult to maintain, calling for ongoing updates and repairs that drive up operational costs. Finally, several stakeholders have questioned the welfare implications of such technology, which raises ethical considerations concerning animal handling. Despite these limits, there has been continuous research into making IoT systems more efficient, affordable, and trustworthy, driven by technology improvements and a rising awareness of animal protection. The following Table 2.2 further

TABLE 2.1
Summarizing the Traditional Crop Protection Techniques and Their Limitations [9, 10]

Crop Protection Technique	Description	Effectiveness	Limitations
Barbed wire fencing	Fencing made from barbed wire to deter animals from entering crop areas.	Effective against large animals such as cattle and deer.	• Requires regular maintenance due to rust and breakage. • Not effective against smaller animals such as rodents. • Time-consuming and labor-intensive installation.
Timber fencing	Wooden fences are used for crop protection from large animals.	Good for deterring large animals (e.g., deer, wild boar).	• High cost for materials and installation. • Can rot, warp, or be damaged by pests (e.g., termites). • Needs regular repairs and upkeep.
Chemical pesticides	Chemicals applied to crops to eliminate pests, insects, and diseases.	Highly effective against pests and insects.	• Harmful to the environment and non-target organisms. • Health risks due to pesticide residues. • Can lead to pest resistance over time.
Physical barriers (e.g., netting)	Mesh or other barriers used to prevent animals from accessing crops.	Effective for small animals and birds.	• Expensive to install on a large scale. • Can degrade over time, requiring replacement. • Limited coverage for larger pests.
Trapping	Traps set to capture or kill pests and small animals.	Effective for localized pest control.	• Requires regular checking and resetting of traps. • Ethical concerns related to animal welfare. • Not scalable for large fields.
Crop rotation and diversification	Rotating crops or diversifying them to manage pests and prevent soil depletion.	Helps in pest control and maintaining soil health.	• Requires careful planning and knowledge of pest behavior. • Not always practical for large commercial farming.
Manual harvesting and pest removal	Manually removing pests or harvesting crops early to prevent damage.	Effective in small-scale farming.	• Very labor-intensive. • Limited applicability for large-scale farming.
Firebreaks and controlled burns	Creating firebreaks or controlled burns to manage wildfires and pests.	Helps in reducing wildfire risks and controlling certain pests.	• Risk of uncontrolled burns. • Can degrade soil quality and harm surrounding vegetation.

TABLE 2.2
Comparison of Challenges Faced by Different Technologies Used for Wildlife Security [12]

Technology	Challenges	Cost and Scalability	Maintenance and Durability	Environmental Impact	Ethical Concerns
IoT-integrated fencing	**High Setup Costs, Power Supply Issues,** Poor **Connectivity Limitations** **Environmental Damage**: Weather and wildlife may impair system performance. **Cybersecurity Vulnerabilities**: Potential for hacking and unauthorized access.	**High Initial Costs**: Substantial setup costs for infrastructure and devices. • **Scalability**: Expanding the system to larger areas can be expensive and logistically complex.	**Frequent Maintenance**: Need for continuous monitoring and repair of components. **Limited Durability**: Vulnerable to weather and wildlife interference.	**Potential Environmental Harm**: The environmental footprint of production and power use of IoT devices. • **Wildlife Disturbance**: Interaction with animals can cause behavioral changes.	**Potential Stress on Wildlife**: Prolonged exposure to electronic systems could stress animals. **Humans vs. Wildlife**: Ethical questions about technology replacing natural monitoring methods.
Traditional fencing (e.g., barbed wire, timber fencing)	**Frequent Repairs**: Regular upkeep and maintenance due to weather damage. **Durability Issues**: Timber fencing decays over time, and barbed wire is easily damaged. **Limited Scalability**: Expanding the fencing system to cover large areas is inefficient and costly.	• **Cost-Effective**: Lower initial cost compared to newer technologies, but requires ongoing maintenance costs. • **Scalability Issues**: Expanding over large areas can be expensive and ineffective.	• **High Maintenance**: Requires constant repairs to withstand harsh weather. • **Durability Problems**: Can easily decay or be damaged, reducing the fence's lifespan.	• **Environmental Impact**: Timber can contribute to deforestation, and barbed wire can pose risks to animals.	• **Risk to Animals**: Barbed wire can injure or harm animals, creating ethical concerns.

(Continued)

TABLE 2.2 (CONTINUED)
Comparison of Challenges Faced by Different Technologies Used for Wildlife Security [12]

Technology	Challenges	Cost and Scalability	Maintenance and Durability	Environmental Impact	Ethical Concerns
Drone surveillance	**High Operational Costs**: Expensive deployment and maintenance costs. **Limited Battery Life**: Drones have a finite range and flight time. **Weather Sensitivity**: Susceptible to adverse weather conditions. **Regulatory Restrictions**: Legal and airspace regulations pose challenges. **Privacy Concerns**: Potential to infringe on privacy in populated areas.	• **Expensive Operational Costs**: Drones require high initial investment and ongoing expenses for operations and maintenance.	• **Frequent Maintenance**: Drones need regular checks and maintenance for optimal performance. • **Battery Life Limitations**: Short operational time limits use.	• **Environmental Impact**: Drone use may have a small environmental footprint but contributes to noise pollution and carbon emissions.	• **Privacy Issues**: Drones might invade privacy in areas where humans live or work. • **Stress to Animals**: The presence of drones may disturb or stress wildlife, especially near nests or dens.
Camera traps and motion sensors	• **Limited Detection Range**: Requires careful placement to cover large areas effectively. • **High Maintenance**: Frequent maintenance of sensors, cameras, and batteries. • **Environmental Factors**: Vulnerable to weather conditions and animal interference. • **Limited Coverage**: Blind spots are common in large areas. • **Data Management**: Handling large amounts of image/video data can be overwhelming and inefficient.	• **Affordable**: Less expensive than advanced technologies, but still requires upkeep costs. • **Scalable**: Easier to scale, though not as effective for vast areas.	• **High Maintenance**: Regular maintenance of equipment is necessary to ensure reliability. • **Vulnerable to Environmental Damage**: Cameras can be damaged by wildlife or extreme weather.	• **Environmental Impact**: Minimal environmental impact, but cameras require energy, which may have a carbon footprint.	• **Animal Welfare Concerns**: Cameras might disrupt animal behavior, especially during breeding or sensitive periods.

(Continued)

TABLE 2.2 (CONTINUED)
Comparison of Challenges Faced by Different Technologies Used for Wildlife Security [12]

Technology	Challenges	Cost and Scalability	Maintenance and Durability	Environmental Impact	Ethical Concerns
Acoustic deterrents	**Effectiveness Variability**: Varies across species, with some animals unaffected. **Limited Range**: Less effective over larger areas. **Environmental Interference**: Noise pollution from weather or other factors can reduce effectiveness. **Ethical Concerns**: Prolonged noise exposure could stress animals. **Maintenance**: Requires regular checking and adjustments for optimal operation.	**Low Initial Cost**: Acoustic deterrents are relatively inexpensive to install. **Scalability**: Can be expanded over larger areas with relatively low additional cost.	**Low Maintenance**: Requires less maintenance compared to physical fencing or advanced monitoring systems.	**Minimal Environmental Impact**: Acoustic deterrents have a lower environmental footprint compared to other technologies.	**Animal Welfare Issues**: Can cause stress or displacement in animals, potentially altering their natural behaviors.
Electric fencing	**High Setup and Operational Costs**: Significant cost for equipment and ongoing electricity usage. **Maintenance**: Requires continuous monitoring and upkeep to prevent failure. **Animal Welfare Concerns**: The system can harm animals that come into contact with the electrified components. **Environmental Sensitivity**: Vulnerable to environmental factors such as heavy rainfall and plant growth, which may short-circuit the system.	**Expensive Setup**: High initial costs for installation and ongoing energy costs. **Scalable**: Expanding electric fencing to larger areas is costly and requires more resources.	**Frequent Maintenance**: Needs regular inspections and repairs, especially during extreme weather conditions. **Susceptible to Weather**: Can malfunction if exposed to severe weather conditions or plant growth.	**Environmental Impact**: Potential environmental harm due to power consumption and disruption of ecosystems.	**Animal Welfare**: Contact with the electric fence can harm animals, leading to ethical concerns.

shows the comparison of challenges faced by different technologies used for wildlife security.

This wildlife security technology comparison shows various advantages and downsides. While the old-school fence provides an extremely low-tech, low-cost approach [14], the newer, IoT-integrated fences represent a data-gathering solution with real-time monitoring capabilities, albeit with high upfront costs, possible connectivity issues, and potential for high replacement/maintenance requirements [15]. Conventional fencing is cost-effective but limited by maintenance demands and impracticality over large expanses and does not serve the goals of contemporary wildlife management. However, despite the potential of drone surveillance to offer accurate observation from the air, it is a costly tool, and this, coupled with a limited battery life and a sensitivity to weather, limits its practical application. Camera traps and motion sensors may work for catchment range, but their point nature makes them more of a maintenance rollout. Although acoustic deterrents are effective against certain species [13–15], they have limited range and limited success in varied environments. Electric fences are often used but are very expensive to run and can be affected by external environmental conditions such as electricity failure. Each technology and tool set brings its own benefits versus many trade-offs, and the final solution depends on the unique needs and constraints of the wildlife manager.

2.4 CONCLUSION

To be able to properly grow crops, it is important to protect them from animals. Advancements in IoT and machine learning technology can help accomplish this. This work covers Raspberry Pi CPU, WiFi module, CNN, and ML methods. This technique has superiority over R-CNN in terms of computation time, accuracy, and effectiveness. In the future, it can be a mobile, user-friendly app-based paradigm. Central to this strategy is the use of recovery zones, which are areas that provide refuge and allow populations to move away from farmlands, further encouraging coexistence with agricultural species while protecting biodiversity. Its modular and scalable architecture makes the system available to different agricultural groups, even to the least privileged groups. Flexible implementation options, from mild to grand, give versatility to farms of all sizes and financial capabilities. The IoT bot's real-time monitoring and reporting functions enable ongoing system optimization, adapting to farmer interactions and field circumstances. Future research will enhance system efficacy, scalability, and long-term impact through major projects. To ensure year-round dependability, the system will undergo multi-season field assessments to evaluate its adaptability to several climatic circumstances, including monsoon, drought, and winter.

REFERENCES

1. Jakes, A.F., Jones, P.F., Paige, L.C., Seidler, R.G. and Huijser, M.P. 2018. A fence runs through it: A call for greater attention to the influence of fences on wildlife and ecosystems. *Biological Conservation*, *227*, pp. 310–318.

2. Burudi, J.W., Tormáné Kovács, E. and Katona, K. 2025. Wildlife fences to mitigate human–wildlife conflicts in Africa: a literature analysis. *Diversity*, *17*(2), p. 87.
3. Abed, N., Murgun, R., Deldari, A., Sankarannair, S. and Ramesh, M.V. 2025. IoT and AI-driven solutions for human-wildlife conflict: Advancing sustainable agriculture and biodiversity conservation. *Smart Agricultural Technology*, *10*, p. 100829.
4. Geisser, H. and Reyer, H.U. 2004. Efficacy of hunting, feeding, and fencing to reduce crop damage by wild boars. *The Journal of Wildlife Management*, *68*(4), pp. 939–946.
5. Singh, P., Chaudhary, M., Saini, N. and Bhatt, R. 2024. WSN application for crop protection to divert animal intrusions in the agricultural land.
6. Tsebesebe, N.T., Mpofu, K., Sivarasu, S. and Mthunzi-Kufa, P. 2025. Arduino-based devices in healthcare and environmental monitoring. *Discover Internet of Things*, *5*(1), pp. 1–31.
7. Enoh, M.K.E., Ahmed, F., Muhammad, T., Yves, I. and Aslam, F. 2023. Navigating utopian futures. *AJPO Journals USA LLC.*
8. Lalitha, R., Hariharan, G. and Lokesh, N., 2020. Tracking the Covid zones through geo-fencing technique. *International Journal of Pervasive Computing and Communications*, *16*(5), pp. 409–417.
9. Zheng, J. and Xu, Y. 2023. A review: development of plant protection methods and advances in pesticide application technology in agro-forestry production. *Agriculture*, *13*(11), p. 2165.
10. Zhou, W., Arcot, Y., Medina, R.F., Bernal, J., Cisneros-Zevallos, L. and Akbulut, M.E. 2024. Integrated pest management: an update on the sustainability approach to crop protection. *ACS omega*, *9*(40), pp. 41130–41147.
11. Karunarathne, S.M., Dray, M., Popov, L., Butler, M., Pennington, C. and Angelopoulos, C.M. 2020. A technological framework for data-driven IoT systems: application on landslide monitoring. *Computer Communications*, *154*, pp. 298–312.
12. Dourish, P., Grinter, R.E., Delgado De La Flor, J. and Joseph, M. 2004. Security in the wild: user strategies for managing security as an everyday, practical problem. *Personal and Ubiquitous Computing*, *8*(6), pp. 391–401.
13. Lahoz-Monfort, J.J. and Magrath, M.J. 2021. A comprehensive overview of technologies for species and habitat monitoring and conservation. *BioScience*, *71*(10), pp. 1038–1062.
14. Javed, S., Higuchi, H., Nagendran, M. and Takekawa, J.Y. 2003. Satellite telemetry and wildlife studies in India: advantages, options and challenges. *Current Science*, *85*(10), pp. 1439–1443.
15. Saha, H.N., Roy, R., Chakraborty, M. and Sarkar, C., 2021. IoT-enabled agricultural system application, challenges and security issues. *Agricultural Informatics: Automation using the IOT and Machine Learning*, pp. 223–247.

3 Ciphering Authenticity
Unveiling Signature Forgery with CNN

Kalyan Chatterjee, Subhasish Roy, and Vanisha Buragohain

3.1 INTRODUCTION

Despite the evolution of digital signatures, the reliance on handwritten signatures persists as a common form of authentication in today's modern age, verifying identities in diverse areas, including, but not limited to, financial transactions, legal documents, official records, and institutional approvals. But the evolution of forgery techniques has gradually eliminated the credibility of this traditional verification approach and, thus, constitutes a danger to security and trust [1]. This chapter describes how a solid-policy-based system using CNNs is developed and deployed with the purpose of identifying forged handwritten signatures [2]. Specifically, CNNs, a type of in-depth learning algorithm that is powerful for solving image-based problems, hold great promise as a way to further secure signature-based authentication [3, 4]. Using CNNs allows the system to accurately evaluate signatures and identify forged ones, thus proving to be an invaluable asset in the fight against fraud and ensuring the integrity of authentication mechanisms [4–7]. The main goal is to increase the security of signature-based authentication systems, especially for offline signatures, where the signature is buffered and evaluated once the signing is finished. It applies techniques of machine learning and CNNs to classify, recognize, and distinguish between genuine and fake signatures. We prioritize the use of TensorFlow, an open-source machine learning framework, for designing a CNN model for the implementation. Moreover, the system has more complex feature extraction techniques based on different geometric characteristics that are extracted from the signature image dataset, allowing it to detect subtle differences and distinguish between real signatures and forgeries.

In this study, we have attempted to break the complexity of a signature forgery into four basic types. Simulation forgery is when a forger tries to mimic a real signature; these are of lower quality than true signatures but can differ depending on skill or practice. Unknown forgery or blind forgery is the type of forgery in which a forged signature is copied without using a model, and the perpetrator's handwriting can generally be observed because the forger's handwriting becomes easier to

DOI: 10.1201/9781003567585-3

identify. This kind of testamentary forgery is perpetrated by tracing the lines of a genuine signature, usually by means of backlighting or strong impression, which is hard to detect short of access to the original document.

3.1.1 Objective

We aim to enhance the security of handwritten signature verification systems by introducing a strong counterfeit signature detecting model. This method concerns developing a database gathering both real and fake signatures for training and testing a convolutional neural network (CNN) model. Through the use of deep learning and dedicated CNN layers for feature extraction, the system is able to recognize complex patterns and slight variations in handwritten signatures. Besides, advanced pre-processing techniques are also employed to enhance the quality of the image and models. The overarching aspiration is to design a more accurate and robust system for recognizing genuine and forged signatures from the authors.

3.1.2 Problem Description

In the area of signature forgery, accuracy is often more important than fluency because forgers tend to focus on mimicking the most important characteristics of a signature, as opposed to the signature as a whole. A good verification system should have one main aim—to do as little false positive (accepting a forgery as real) and as little false negative (denying a real signature as a fake) as possible. Accurate signature verification is a rare and difficult task, and human examiners tend to use several key features to distinguish between a fraudulent and a real signature.

Detection of forgery in handwriting and signatures is a process of examining several features that could show abnormalities and are not nature's creations. Obvious ones would be shaky handwriting that betrays a lack of confidence in the fraudster and unnatural picks up of the pen that break the flow of the writing. Retouch marks or overwrites can be signs that the counterfeiter was trying his best to get the signature right. Disproportionate letters and irregular shape or size of the signature and a variety in the slant or the angles would also indicate that it could be contrived. Highly similar signatures from one card to another, for example, might indicate tracing or copying, while unusual speed or pressure, be it too fast, too slow, too light, or too heavy, could be a sign of hesitation or forced execution.

3.2 LITERARY SURVEY

Building on research in signature forgery detection, this project has provided a number of insights and inspirations:

Detection of Handwritten Signature Forgery Using CNN: The way of handprint verification and forgery identification is discussed, focusing specifically on the use of CNN algorithms to detect handwritten signature and drawing the eyes at any cost to it. It explores classification algorithm

Hand verification, which is divided into static (offline) and dynamic (online) [8], emphasizing the use of deep learning algorithms, in particular CNNs, namely in propaganda model building, that reveal unique outline features from which a classification can be trained to separate genuine signatures from false samples [9].

From Text to Signature: Knowledge Transfer for Efficient Deep Feature Learning on Offline Signature Verification focuses on off-line signature verification (OSV) with deep CNNs to learn features [7, 10]. The authors propose a knowledge transfer strategy between a CNN model pre-schooled in handwritten text writer recognition and its first steps to shortened distance learning speed in OSV [11]. This chapter introduces an offline signature [12] verification method that makes use of Convolutional Siamese Networks. The architecture consists of two symmetrical but oppositely oriented convolutional networks with shared weights, one processing each part image in overlapping windows, thus forming a feature map to facilitate signature comparison [13].

3.3 APPLICATIONS

To ensure that a signature [14] has not been forged, detection is necessary in many fields. This ensures that people are signing their own names (and all the attendant rights with respect to those names), and it also provides a layer of protection against fraudulent activities.

3.3.1 Banking and Financial: Signature Forgery Detection

In banking and financial services, the process of signature verification is important for ensuring fraud deterrence and the security of the institution. Documents such as contracts and agreements require original signatures to evidence their authenticity, and the ability to detect forgery is accordingly of crucial importance. Signature forgery is the process of making, adapting, or imitating a signature without there being fraudulent intent. Early detection of forged signatures is crucial to minimize fraud in applications such as issuance of drivers' licenses or passports. As digital transactions become more prevalent, the modification of such signature tests when used in an online context is necessary for one-party contracts and e-commerce, for example. In art, as in many other fields, authenticity of the signature ensures the worth and trust in the collectible. In forensic document examination, a document submitted as evidence in court must be verified through signature analysis, and authentic signatures are also required for notary services; this places a high demand on the robustness of the detection in these domains [8, 10, 12].

3.3.2 CNN Architecture

A CNN is a specific type of artificial neural network that uses a mathematical operation known as convolutions, which, in turn, makes it ideal for working with data having a known grid topology, such as images (2D grid of pixels). CNNs are designed based on the architecture of the animal visual cortex, which is structured into several layers that consist of small groups of neurons that each respond to different areas of the visual field and transmit processed information to other layers for additional processing. A CNN has a complete hierarchy of neurons—however, each CNN unit only operates in its receptive field (similar to the visual field of a neuron in the visual cortex). In CNN, the layers are arranged hierarchically where the first layers identify simple patterns such as edges and lines, while deeper layers are responsible for identifying more complex patterns such as shapes and objects. The hierarchical structure allows CNNs to extract high-level features from images, making them particularly effective for image recognition and classification tasks.

3.3.3 Convolutional Layer

The convolutional layer forms the basic unit of a CNN. It highlights the features of the input image by computing the dot product between a filter (called kernel) and a small part of the input image. The input image is a matrix of pixels, and the filter is also a smaller matrix than the input image and detects the specific features in the image, such as edges or corners. The convolution operation is performed by moving a filter across the input image and calculating a dot product at each position. Each of these results of the dot product will give a single value or a number, and collectively they will form up a new matrix called the "feature map."

3.3.4 Fully Connected Layer

A fully connected layer is a layer in which all the neurons are connected to all the neurons in the previous layer. Essentially, this layer takes as input from the last pooling layer or the last convolutional layer and linearly transforms the input data. The output from a fully connected layer, then, is passed through an activation function to add non-linearity to the model. All together, these layers make ConvNets a powerful and widely used framework for image classification, segmentation, and other tasks that involve visual data. Then, they are combined into some higher-level representation that is used to make classification (or other) decisions.

3.3.5 Output Layer

The output layer is the last layer of a CNN architecture. Then it outputs the final result of the model, which might be a class label, a probability distribution across classes, or a continuous value, depending on the problem at hand. Both of these are defined by the classification problem; the number of neurons in the output layer will be equal to the total number of classes. The output layer has one neuron for binary class problems, and for multi-class classification problems, the output layer has one neuron for each class. The output layer uses an activation function depending on the task. The most common activation function used for binary classification is the sigmoid activation function, and for multi-class classification, the softmax activation function is used.

3.3.6 Activation Functions

Activation functions are propelling non-linearity to the model, with the help of which the model is able to learn complex patterns and relationships in the data. All of them are applied to the output of each neuron in the convolutional and fully connected layers.

3.4 METHODOLOGY

This section elaborates on the methodology employed in developing and implementing the CNN-based signature forgery detection system. The methodology encompasses several key stages, including data acquisition, preprocessing, model construction, training, and evaluation.

3.4.1 Data Acquisition

In every machine learning project the starting point is having a good dataset. A handwritten signature dataset was prepared meticulously for this project. The dataset contains a total of 120 signature [15] images for each person, with 60 genuine and 60 forged signatures. Images were sourced from different people to maximize the variety in writing styles and forgery methods (Figure 3.1). Each image in the dataset is an RGB (red-green-blue) image, ensuring that the color information of the signature is not lost/affected.

FIGURE 3.1 Real signature dataset.

3.4.2 Preprocessing

Preprocessing is an essential process for the preparation of signature images for effective feature extraction and model training. The following are the preprocessing steps:

1. **Resizing**: All signature images were resized to a consistent 224 × 224 pixel size. This ensures consistency in the data and minimizes the computations during model training.
2. **Convert the images to Grayscale**: To reduce the dimensionality of the data, the captured RGB images were converted to grayscale. Since the features of the signatures are mostly defined by the intensity variations, grayscale images contain all the necessary information.
3. **Normalization**: Pixel intensities were scaled to values between 0 and 1. This process provides stability during the training process and also increases the rate of convergence of the model.
4. **Background Removal**: Sometimes, background removal methods were used to separate the signature from its background, enabling more effective feature extraction by the CNN model.

3.4.3 Model Construction

At the heart of the signature forgery detection system is the CNN model. We built the CNN model using the Keras library, a high-level API to build and train deep learning models. The model architecture consists of different layers, each designed to fulfill a specific task:

1. **Convolutional Layers**: These layers are key extracts from input signature images. They use learnable filters that slide over the image and recognize features such as edges, corners, and curves. Each convolutional layer produces a set of feature maps that indicate the presence and strength of certain features in the image.
2. **Pooling Layers**: These layers downsample the feature maps to reduce the height and width of the representation. Downsampling achieves multiple advantages, such as computational cost reduction, translation invariance, and overfitting avoidance.
3. **Fully Connected Layers**: These layers will learn some of the complex relationships from the features seen in the convolutional and pooling layers. They bring the features together in a higher-level representation suitable for classification.
4. **Output Layer**: This layer is the final layer of our model and gives us the output, where the output is a binary classification of whether the input is a genuine signature or a forged signature.

3.4.4 Training

The CNN model constructed was trained with the processed signature dataset. Based on the differences between the predictions from the model and the ground truth labels, the model learns the difference between the original and the forged signature by modifying its parameters.

The training was performed with Adam, an adaptive optimization algorithm that can compute different learning rates for different parameters. This loss function can be measured with respect to the output of the model and its ground truth label, hence the use of binary cross-entropy. However, to avoid overfitting to the training set, early stopping was used.

Early stopping can also be used to prevent this problem by monitoring the validation loss over time and stopping training when the validation loss does not improve after *n* epochs.

3.5 RESULTS AND DISCUSSION

This section presents the results obtained from the implementation of the CNN-based signature forgery detection system and provides detailed discussions on the model's performance, interpretations of the results, and analysis of various signature comparisons.

3.5.1 Overall Performance

A CNN was trained on the signature dataset preprocessed dataset, returning a 97% accuracy in distinguishing genuine and forged signatures across the test dataset. This level of accuracy shows how successful the model is at discerning real from fake signatures and proves proof-of-concept for real-world applications in industries with signature verification requirements. This model performs vastly well, owing to a couple of things:

- **Strong Dataset**: The large and varied dataset for training and evaluation allowed the model to see diverse writing styles and forgery methods, which helped it learn and generalize well.
- **Preprocessing Techniques**: The rigorous preprocessing steps applied, such as resizing, normalization, and data augmentation, not only enhanced the quality of the signature images but also made the model more competent in extracting relevant features from the signatures.
- **CNN Architecture**: The complex architecture of CNN with its convolutional, pooling, and fully connected layers allowed us to learn the fine edges and patterns that differentiated a real signature from a replicate.

3.5.2 Training and Validation Performance

The model is experiencing the learning process in training, which is illustrated in Figure 3.2. The graph displays the training and validation accuracy and loss over the training epochs. The line in the graph at 0.70, denotes training accuracy, which, in fact, increases progressively until it reaches ~100%. This means that the model is learning the patterns and features of the training data properly. The line in the graph below 0.70 represents the validation accuracy. It fluctuates initially but stabilizes just below 97% Such an oscillation is normal, as the algorithm faces unseen validation data and adapts its parameters to generalize better.

In Figure 3.2, the model is learning from the training data, generalizing well to the validation data, and achieving high accuracy without overfitting.

3.5.3 Signature Comparisons and Analysis

Figure 3.3 showcases comparisons between forged and genuine signatures, along with the model's predictions and accuracy for each case.

Figure 3.3 shows the comparison of forged signature (left one) versus genuine signature (right one) used at testing. The forged signature was predicted by the model with an accuracy of 97.22%. This shows that the model can catch very small differences between real and fake that they are visually similar.

In Figure 3.4 we show another instance of a forged signature and a genuine signature. Yet another demonstration of the effectiveness of the model was that it predicted the forged signature with an accuracy of 96.78%.

As an example, the model also succeeded since it has accurately evaluated a forged signature (Figure 3.5 a) in relation to the genuine signature (Figure 3.5 b), with an accuracy of 97.22% (Figure 3.5 c). This demonstrates the robustness and reliability of the model in detecting forgeries, as can be seen from the high accuracy across varying samples of the signature.

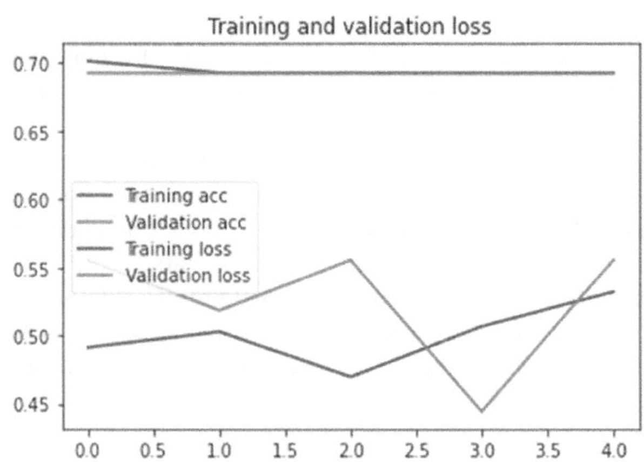

FIGURE 3.2 Training and validation accuracy and loss.

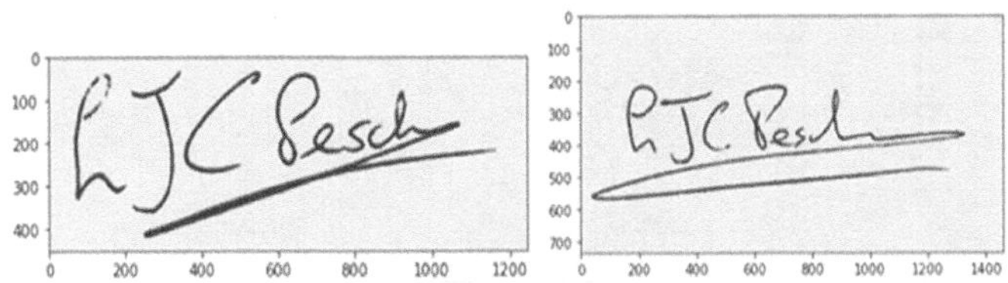

FIGURE 3.3 Forged and genuine signature comparison with accuracy.

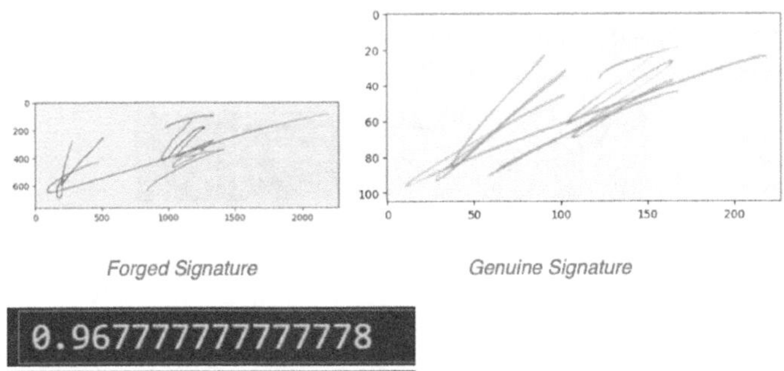

FIGURE 3.4 Forged and genuine signature comparison with accuracy.

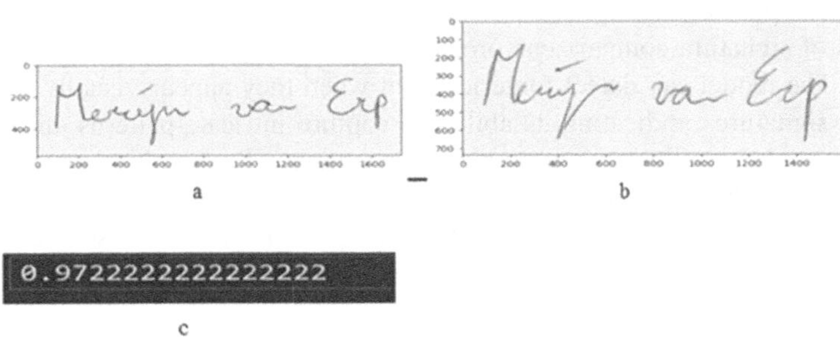

FIGURE 3.5 Forged and genuine signature comparison with accuracy.

Figure 3.6 exhibits a forged signature and an authentic signature. On the other hand, it attained 98.78% accuracy by properly categorizing the forger sign, revealing its capacity to identify slight variations that may suggest forgery. The model detected the forged signature with 97.78% accuracy. Moreover, the model's performance over a variety of signature styles and forgery attempts also serves as an evidence for the model's robustness.

The model was able to accurately detect the forgery with a 97.22% accuracy (Figure 3.7). With the consistently high accuracy achieved in all web samples tested, the model proves to be a reliable and viable real-world solution to the issue of signature forgery detection.

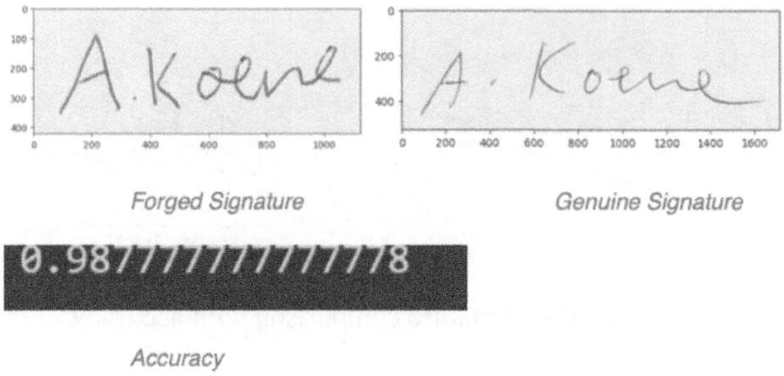

FIGURE 3.6 Forged and genuine signature comparison with accuracy.

FIGURE 3.7 Accuracy.

3.6 DISCUSSION

The results presented in this section demonstrate the effectiveness of the CNN-based signature forgery detection system. The model's high accuracy, consistent performance across various signature styles and forgery attempts, and ability to identify subtle discrepancies highlight its potential for real-world applications. The detailed analysis of signature comparisons provides further insights into the model's capabilities. The model can detect forgeries even when they appear visually similar to genuine signatures, indicating its ability to capture intricate patterns and nuances that may not be readily apparent to human examiners. The results also suggest that the model is robust and reliable, as it consistently achieves high accuracy across different signature samples and forgery techniques. This robustness is crucial for real-world applications where the model needs to handle a wide range of scenarios and variations.

3.7 CONCLUSION

In this chapter, we have presented the evolution and construction of a strong signature forgery detector using a CNN. Using deep learning, the system can recognize and differentiate between real and fake signatures, a vital security issue in many industries. These include the precise construction and cleaning of a diverse signature dataset, the careful architecture of a CNN, the tuning of an appropriate training method, and a full performance evaluation of the model.

As the results obtained in this chapter show, the proposed CNN-based system successfully detects signature forgeries with very high rates of true positive detections. Its heightened sensitivity to variances, consistent performance across diverse signature models and forgery techniques, and robustness to noise continue to underscore its potential for practical significance. Future research in the domains mentioned here would help further improve the accuracy, security, and reliability of these systems, leading to a more secure and trustworthy authentication ecosystem.

REFERENCES

1. J. M. Vargas, M. A. Ferrer, C. M. Travieso, and J. B. Alonso, "Off-line signature verification based on grey level information using texture features," *Pattern Recognition*, vol. 44, no. 2, pp. 375–385, 2011.
2. M. R. Alaei, P. Nagabhushan, and U. Pal, "A new two-stage scheme for offline handwritten signature verification," *IEEE Trans. on Information Forensics and Security*, vol. 11, no. 9, pp. 2045–2054, 2016.
3. D. Coetzer, B. M. Herbst, and J. A. du Preez, "Offline signature verification using the discrete Radon transform and a hidden Markov model," *EURASIP Journal on Advances in Signal Processing*, vol. 2009, no. 1, p. 382391, 2009.
4. A. Gattal, A. Belhadef, and K. Mesghouni, "A hybrid approach to offline signature verification using deep learning and hand-crafted features," *Neural Computing and Applications*, vol. 33, no. 13, pp. 8159–8173, 2021.
5. S. A. R. S. Al-Hammadi and A. K. Nandi, "Deep learning for authentication and identification of handwritten signatures," *IEEE Trans. on Cybernetics*, vol. 51, no. 5, pp. 2538–2551, 2021.
6. X. Yang, C. Li, and Y. Liu, "Handwritten signature verification using siamese neural networks," *Pattern Recognition*, vol. 96, p. 106972, 2019.
7. R. Plamondon and G. Lorette, "Automatic signature verification and writer identification—The state of the art," *Pattern Recognition*, vol. 22, no. 2, pp. 107–131, 1989.
8. A. K. Jain, F. D. Griess, and S. D. Connell, "On-line signature verification," *Pattern Recognition*, vol. 35, no. 12, pp. 2963–2972, 2002.
9. N. Otsu, "A threshold selection method from gray-level histograms," *IEEE Transactions on Systems, Man, and Cybernetics*, vol. 9, no. 1, pp. 62–66, 1979.
10. Y. Xiong and T. Long, "An efficient CNN model for offline signature verification with class imbalance handling," *IEEE Access*, vol. 8, pp. 189175–189185, 2020.
11. K. Plamondon, G. Bernard, and L. Richard, "Survey of biometric handwriting authentication systems," *IEEE Transactions on Biometrics*, vol. 6, no. 4, pp. 459–469, 2021.
12. M. N. As'ad, M. H. Marhaban, and M. N. Ismail, "Feature-based offline signature verification using Convolutional Neural Networks," *IEEE Access*, vol. 8, pp. 50941–50952, 2020.
13. C. Vivaracho-Pascual and J. Ruiz-Del-Solar, "Deep learning applied to biometric authentication," *Expert Systems with Applications*, vol. 162, p. 113722, 2020.
14. A. H. Soliman and H. A. Khalil, "Signature verification using ResNet-based deep learning model," *Neural Computing and Applications*, vol. 33, no. 16, pp. 10005–10017, 2021.
15. M. R. Haque and S. B. Menhaj, "Deep signature analysis for offline verification," *IEEE Transactions on Image Processing*, vol. 32, no. 1, pp. 67–78, 2023.

4 AI Governance and Cybersecurity

A Legal Framework for Safeguarding the Future of Machine Intelligence

Shashwata Sahu and Navonita Mallick

4.1 INTRODUCTION

AI has mushroomed at light speed and transformed entire sectors that are nowadays critical components in healthcare, finance, defense, and even national security sectors. The former's capabilities at data processing, plus its capacity in decision-making by efficiency and effectivity, made such a wonder applicable in most present-day operations. It also leads to vulnerability and reliance on AI (Sheehan et al., 2023), as the AI system works harder and deals with sensitive data (Bajwa et al., 2021). Therefore, it has become a center for cyberattacks, data breaches, and manipulation. These areas are more risky since critical infrastructure includes the power grid, defense systems, and self-driving vehicles. If found vulnerable, such areas would have catastrophic results for the concerned country (Islam et al., 2023). So, with AI, the requirement of strong legal structures that regulate the governance and cybersecurity of AI will be in tandem with its growth. The regulatory framework remains scattered and uneven across jurisdictions, irrespective of the increasing speed of AI development. This makes it challenging for cybersecurity (Binhammad et al. 2024), accountability, and ethics in AI implementation without clear laws that can be enforced regarding governance in AI. Some regions, such as the European Union (EU), have adopted regulations such as the General Data Protection Regulation (GDPR) and the Cybersecurity Act. Different industries have industry-specific regulations (European Parliament, Directorate General for Parliamentary Research Services, 2020). This discrepancy in regulation strategy is turned into loopholes for exploitation by malicious actors who substitute with deepfake manipulation, AI-driven fraud, or algorithmic bias. More importantly, a lack of international cooperation in AI governance undermines efforts to combat transnational cyber challenges. When moving toward AI, say, call on policymakers and regulators to deliver robust legal frameworks guaranteeing AI

 DOI: 10.1201/9781003567585-4

technologies are properly and safely developed and messaging with transparency and ethics (Kavanagh, 2019).

4.2 RESEARCH OBJECTIVES

The chapter analyzes current legal frameworks that regulate AI and cybersecurity (Christensen et al., 2017), discussing their effectiveness vis-à-vis emerging technological challenges. Because of the high-speed rate of AI integration into vital domains, it is time, with so many AI manifestations circulating, to gauge what is already on the books regarding security, accountability, and ethical deployment. It attempts to show gaps and contradictions in national and international regulatory frameworks and highlight shortcomings, such as the fragmentation of governance, weak implementation, and evolving cyber threats (Alevizos, L. et al., 2025). Ultimately, this thesis seeks to facilitate a harmonious coexistence of technological advancement and legal safeguards in a future cyber hub of modernistic frameworks.

4.2.1 Research Methodology

The chapter adopts a doctrinal approach, analyzing relevant primary and secondary sources of law to identify applicable statutes, international conventions, case laws, and pertinent literature. This gives us an accurate picture of what is happening and what already works in AI and cybersecurity. This method has been employed to understand how jurisdictional framing of the approaches taken concerning the EU's GDPR and the United States' AI and cybersecurity, for example, addresses these questions. These are illustrated by integrating international considerations through the reports of global organizations such as the United Nations and the World Economic Forum.

4.3 CONCEPTUAL AND THEORETICAL FRAMEWORK OF AI GOVERNANCE AND CYBERSECURITY

AI governance refers to policies, regulations, and ethical guidelines designed to ensure the safe, secure, and responsible development, deployment, and use of artificial intelligence technologies. As AI becomes entrenched within healthcare and finance systems via national security and law enforcement, there is a growing need for transparent governance. Unregulated AI could entrench bias, compromise security, and act with impunity, problems that pose severe legal and ethical dilemmas (Naman Yadav, 2025). Now, such balancing of innovation with regulation is attempted in AI governance, which aims to design and implement AI systems that work not so much to harm but to help protect human rights and prevent harm to society so that society's welfare improves. Governments and global institutions have, in principle, been open to creating regulatory frameworks such as the EU's GDPR and the proposed EU AI Act that would place legal obligations on developers and users of AI. These frameworks are fragmented, and few countries have comprehensive

AI-specific laws (David et al., 2024). Many jurisdictions adopt a sectoral approach, imposing a different regulatory regime depending on the application industry for the AI technology. In contrast, the rest of the folks call for overarching, unified governance structures that allow consistent AI policies across sectors.

The AI system will produce incorrect or biased outputs if those datasets are hacked or altered. Data poisoning attacks are a type of attack in which deliberately misleading data is input to a training model of an AI and causes the modeled system to yield incorrect predictions or classifications (Kaur et al., 2023). These have highly adverse effects across healthcare and criminal justice domains that use generative AI for diagnostic and risk assessment (Baig et al., 2024) purposes. A key current cyber threat to combat is adversarial attacks, in which hackers tamper with the inputs fed into the AI to trick the system. One of the corner examples of this is in an image recognition AI; even very slight and imperceptible changes in an image are enough to make an AI recognize the wrong thing (Gilbert & Gilbert, 2024). Such attacks represent numerous threats to AI-dependent security platforms, autonomous vehicles, and facial recognition systems concerning bypassing authentication mechanisms or real-world safety risks. AI bias is another significant issue when it comes to cybersecurity and governance. AI algorithms can be biased due to skewed datasets, incorrect model training, or lack of oversight by regulatory bodies. If the historical data collected by these AI systems represented historical data biases, AI would continue to discriminate in hiring and lending, in police arrests, and in healthcare decisions (Girdhar et al., 2023). Governance frameworks should implement a mechanism to audit bias in AI systems so that the technologies operate fairly and equitably.

Ethically, AI governance ensures that AI aligns with social values and human rights. Many ethical frameworks demand AI transparency, accountability, and fairness. Generally, AI transparency refers to the extent to which an AI system can explain the decision-making processes in terms that humans can understand. The black-box problem faces one of the most significant challenges in AI governance when complex machine learning models yield decisions for which even their developers cannot readily explain the rationale behind those decisions (Memarian & Doleck, 2023). For example, transparency laws such as the right to explanation in GDPR require AI-driven decisions to be explainable and justifiable, often on sensitive issues such as hiring and lending or legal adjudication. AI accountability leads to a scenario where deployers of the AI systems, that is, the developers, companies, and governments, take more responsibility. Accountability cannot be readily established if AI systems are deployed and cause harm since no legal liability framework exists. For example, who is to blame in an autonomous vehicle accident? Is it the manufacturer, the software developer, or the vehicle's owner. Legal scholars further promote algorithmic accountability, as decision-making through AI should be checked, validated, and tested on the system's ability not to cause damage (Rodrigues, 2020).

Another commonly accepted theory in the cybersecurity domain that is easily associated with the governance of AI is the Zero-Trust Security Model. It assumes that no user, device, or system can be trusted. All require ongoing authentication and verification. AI security zero-trust frameworks also prevent unauthorized access and help mitigate insider threats to secure data further. AI governance combines security layers, including multifactor authentication, end-to-end encryption, and continuous, systematic monitoring to ensure enhanced cybersecurity against cyberattacks (Kang et al., 2023). One critical cybersecurity concept is AI resilience, which builds a potent and resilient AI system to detect and respond to cybersecurity incidents. Keys to this will be designing AI models resistant to adversarial attacks, implementing self-healing algorithms, and ensuring that AI functions can be secured in adversarial environments. Secondly, any autonomous system where cyber vulnerabilities can lead to loss of life, such as a self-driving car or AI-powered defense system, requires AI resilience (Salem et al., 2024).

4.4 THE CURRENT LEGAL LANDSCAPE OF AI GOVERNANCE AND CYBERSECURITY

Innovative applications of AI are revolutionizing every available industry, from healthcare to finance and defense to governance, creating an urgent need for robust legal frameworks for AI development and deployment. Machine learning and big data-based AI systems can also be vulnerable to cybersecurity, privacy issues, and unethical decision-making (Rashid & Kausik, 2024). Here, we describe relevant legal frameworks on AI governance and cybersecurity, which will be stitched together based on the disparate jurisdictions where you can acquire incomplete information. It thus contrasts regional regulatory models, highlighting drivers of divergence and significant gaps to be filled for a more coherent global governance system.

4.4.1 Evaluation of Existing AI and Cybersecurity Frameworks

Among the most potent custodians of the highest standards for data collection, processing, and storage is the EU's GDPR, a legal framework for data security, particularly for AI, that protects personal kinds of information from unauthorized use and access by requiring strong standards. While initially, GDPR lacked the intention to regulate AI, it serves as one of the most critical EU regulations and directly regulates AI-driven systems, which predominantly rely on a wide range of personal data such as automated decision-making, biometric recognition, and AI-based profiling (Ye et al., 2024). This is one of its strong points, as it proffers significant data protection rights for people, thus the right to be given information, allowing access and objecting to automated decisions. It further requires severe punishment upon defaulting. This would mean that the developers of AI and those organizations handling personal data pay dearly for their malpractices.

The Cybersecurity Act—EU: The EU Cybersecurity Act was enacted in 2019 and aims to enhance the regional cybersecurity frameworks. It encompasses frameworks on AI systems, elevating the status of the European Union Agency for Cybersecurity (ENISA) and providing an EU-wide cybersecurity certification framework to raise the security standards. Its greatest strength is standardizing cybersecurity protocols, which assists AI systems in developing more robust resilience against cyber threats. It also includes the technology vendor certification program that complies with the adopted cybersecurity standards (Almeida, 2025). Nevertheless, with such pros, some serious cons are visible in the Cybersecurity Act, including undefined AI governance; the law is more focused on a broad scope of cybersecurity policy than AI-driven threats such as adversarial attacks and vulnerability through an algorithm. Moreover, the certification programs under this act are voluntarily made (Nihad Hassan, 2025).

The AI Regulation (EU AI Act)—EU: The proposed EU AI Act creates an all-encompassing regulatory framework for AI and categorizes AI systems based on unacceptable, high, and limited risk classifications. The following are some of its significant strengths, including explicit guidelines for applying AI in high-risk systems, such as biometric identification and critical infrastructure. Besides that, the Act requires human supervision in AI decisions, which helps prevent destructive machine decisions that adversely affect someone or the world (Jariwala, 2024).

The United States' Approach—Sectoral and Industry-Specific Regulations: The United States differs from the approach of the EU, for it follows the sectoral approach of AI and cybersecurity governance. In this context, the governing bodies responsible for AI and cybersecurity (such as the Federal Trade Commission (FTC), the Department of Defense (DoD), and the National Institute of Standards and Technology (NIST)) operate within their respective sectors. Such regulations prove burden-free for the development of AI and nurture it to a great extent, which would be a boon, as there would not be any legal hurdles or needless laws hampering such innovations. The sector-specific bodies, which adapt according to the demands and needs of the specific industries dealing with evolving risk contexts, make it possible for American governance to adopt this regulatory approach to AI (Engler, 2023). Regional variations exist in AI security standards (William Simpson, 2023).

4.4.2 Comparative Legal Analysis of AI Governance and Cybersecurity

Table 4.1 outlines the various AI regulatory frameworks, cybersecurity plans, prominent strengths, and shortfalls across the major jurisdictions. It helps illuminate different governance strategies, from the EU's centralized and risk-based model to the United States' sectoral and decentralized approach.

4.4.3 Identified Regulatory Gaps and Challenges

AI governance remains dispersed, and huge differences are found between the models applied in each jurisdiction. The EU prefers risk-based, all-embracing AI governance; however, it applies sector-specific regulations in the United States. China maintains its state-led AI policies, which the government strictly oversees. These different approaches will create legal uncertainty for private-sector businesses operating globally. Other imbalances that increase AI systems' vulnerability are legal loopholes regarding AI-specific cybersecurity measures. Cross-border applications

TABLE 4.1
AI Regulation and Cybersecurity Approaches across Jurisdictions

Jurisdiction	AI Regulation Model	Cybersecurity Approach	Key Strengths	Key Weaknesses
EU (GDPR, AI Act, Cybersecurity Act)	Centralized, risk-based AI regulation	Unified cybersecurity laws across EU nations	Strong data protection and AI risk classification	Compliance burden, slow implementation
US (NIST, FTC, DoD)	Sectoral approach (industry-specific laws)	Decentralized, varies by state and industry	Encourages innovation and market-driven AI development	Lack of federal AI governance, inconsistent cybersecurity standards
China (Cybersecurity Law, AI Ethics)	State-led AI regulation and strict government oversight	High government control over AI cybersecurity	Vigorous enforcement and rapid implementation	Ethical concerns over privacy, government surveillance
India (Proposed Data Protection Bill)	Emerging AI regulatory framework	Limited AI-specific cybersecurity policies	Growing AI governance initiatives	Still under development, lacks enforcement mechanisms

of AI, which frequently entail the processing and movement of data across jurisdictions, raise fundamental questions on cross-border data protection and AI accountability (Meghwal, 2024). Data sovereignty and compliance requirements stemming from private law concepts have arisen in its wake. However, structuring international data-sharing agreements remains a barrier to AI regulatory convergence because they clash with the robust data protection rules of regimes such as the GDPR and the national security policy provisions of regimes such as China's Cybersecurity Law.

4.4.4 Core Principles for Effective AI Governance

Effective AI governance has to depend on some core principles for AI's ethical, secure, and accountable deployment. To avoid potential biases and unfair or otherwise unethical outcomes, transparency in the AI decision-making process is required; hence, these AI models require explainability and audibility so users and regulators know how these decisions come about. Implementing "right to explanation" policies will take transparency to new heights, mandating the developers of algorithms to justify decision-making processes in high-stakes applications such as hiring, lending, and criminal justice (Papagiannidis et al., 2025, Malgieri & Pasquale, 2024).

4.5 PROPOSED LEGAL AND CYBERSECURITY FRAMEWORK FOR AI

At an accelerated pace, as AI technologies further penetrate critical infrastructures, the world's financial systems, and the decision-making process places an increasing urgency for more structured legal cybersecurity guidelines. The identified gaps concerning AI governance or cybersecurity laws across different jurisdictions would need a globally harmonized perspective ensuring security in its development, accountability, and fairness (Crichton et al., 2024).

4.5.1 Establishing AI Regulatory and Compliance Mechanisms

The final rule of responsible AI is the development of regulatory mechanisms, which are dedicated regulatory mechanisms on the aspects of AI governance and compliance. Given that existing rules, including the EU's GDPR and US sectoral law, fall short of addressing AI's unique risks, new governance bodies should be established that fit legal safeguards. Governments need to develop independent AI governance bodies that oversee the compliance of AI systems with ethical and legal standards. These governing bodies would be similar to data protection authorities under the GDPR but with broader powers to regulate AI. They would be instrumental in certifying AI models before deployment to ensure they meet security and ethical thresholds, mitigating risks that AI-driven decision-making could create (Cordella & Gualdi, 2024). Detection audits of bias in AI applications toward industries to enable AI mechanisms to make significantly less biased decisions will also be part of your key duties (Folorunso et al., 2024). AI security laws provide environments for training AI models, encrypt AI-generated data (McKinsey et al., 2023), and guard against adversarial attacks that lead to misalignment in AI decision-making (Mekkawi, 2025). We will create algorithmic accountability laws that hold developers and institutions responsible for malicious harms caused by biased, misleading, or dangerous AI systems and secure the complacent application of AI technology that becomes impossible to operate without the full force of ethical or legal scrutiny. Explainability laws and transparency laws would require recording and disclosing how AI reaches its decisions in criminal justice, healthcare, and other applications, where it directly impacts human lives on a much larger scale (Bartsch et al., 2024). Otherwise,

4.6 IMPLEMENTATION STRATEGIES FOR STRENGTHENING AI SECURITY AND REGULATION

The more AI technologies embedded in essential sectors, including healthcare, finance, defense, and law enforcement, the greater the impetus to heighten effective security measures and regulatory frameworks for these technologies. While the legal and policy recommendations provide a basis for governance for AI, translating those into enforceable practical strategies ensures that its processes are secure, accountable, and fair in deployment (Taeihagh, 2021). This section describes concrete, practical steps to close the regulatory gaps and proposes a hybrid governance model combining legal oversight with industry-driven best practices. Hence, AI is not only secure and ethical but also fosters innovation.

4.6.1 Addressing Key Regulatory Gaps in AI Governance

Given the rapidly evolving landscape, it is unsurprising that there is a lack of consensus around what constitutes responsible AI practice, with definitions of liability, fairness, and transparency often still highly fragmented in existing legal frameworks, even as awareness of the risks of AI continues to grow. To fill these gaps and to constrict the functioning of AI systems to the extent of ethical and legal boundaries, a structured approach/framework is required. Determining who is responsible when an AI system fails, causes damage, or is used by cybercriminals is one of the most pressing issues in AI governance (Birkstedt et al., 2023). Traditional liability doctrines rooted in human decision-making find it challenging to determine who is responsible for AI-driven processes, especially for autonomous systems, other high-risk industries such as self-driving vehicles and medical diagnostics equipment, and AI-powered trading on financial markets. To address this, governments should establish clear liability laws for those who will ultimately be liable when AI harms someone: the developer of the AI, the provider of data, the organization using AI, or all of the above (Vasudevan, n.d.). AI insurance policies require companies using AI systems to have insurance against damages from AI failures or cybersecurity breaches (House of Lords, 2023). Data quality standards must also be developed, and bias detection and audit mechanisms must be used as components of the AI governance frameworks. AI audits should be periodic and mandated by the regulatory authorities (Celi et al., 2022). AI models must undergo bias assessments and fairness tests before deployment, especially in high-risk applications, where biased decision-making adversely affects lives. They should also mandate that organizations provide explainability and documentation and clearly articulated rationales for AI-driven decisions, especially in high-stakes sectors such as hiring practices and criminal sentencing, where fairness is essential to justice and social equity. To advance away from algorithmic discrimination, government and AI regulatory oversight organizations should establish standardized benchmarks for fairness and algorithms to guide AI developers as they build free algorithms and abide by ethical AI standards (Belenguer, 2022). Policymakers can mitigate such risks by instituting bias detection and AI audit mechanisms so that AI is a source of fairness, equity, and inclusion, not a means of perpetuating existing discriminatory practices or disproportionately harming vulnerable populations.

4.6.2 A Hybrid Approach to AI Security and Regulation

Celestial rules of AI governance: Governance of AI cannot be draconian or static. Such a hybrid solution, where regulatory oversight provides a framework, and the industry is given freedom in between its best practices, can create a flexible and adaptive model to keep pace with the evolution of AI. Instead of the excessive burden of implementing regulations that restrict innovation, governments should partner with movers and shakers to jointly develop frameworks of AI security and governance that lead to compliance with regulations without suffocating technological progress (Vedant Sharma, 2024). Part of this plan includes industry-led AI governance councils so that businesses will be self-regulated, provided that all follow the clear lines set by the government, ensuring that the innovation in the AI space happens with compliance and applied ethics in mind. Governments should promote adherence to civilian AI security standards such as ISO/IEC 27001 (for cybersecurity) and ISO/IEC 42001 (for AI management systems) so that AI systems comply with internationally accepted security goals. Another key step would be establishing testing environments known as "sandboxes" so that AI developers can test their security compliance in a controlled environment before being deployed in the real world so that regulators can offer guidance and modifications without the risk of hype resulting in premature deployment of unsafe AI models (Sullivan, 2024).

4.7 CONCLUSION

AI is developing rapidly, bringing potential and challenges in many sectors, such as healthcare, finance, and defense. Although AI-driven systems accelerate processes and enhance decision-making, their adoption in mission-critical systems introduces serious governance and cybersecurity risks. This study reviewed the legal landscape and identified various areas outside AI and cybersecurity regulations. It has also proposed a comprehensive legal and cybersecurity framework and urged transparency, accountability, and international coordination (Yelne et al., 2023). However, the efficacy of these recommendations relies on ongoing coordination and cooperation between governments, private enterprises, and international institutions. Regarding AI governance and cybersecurity laws, the report notes that existing infrastructures are often siloed, with varying limits and regulations on AI in different jurisdictions. The EU's GDPR and AI Act strongly focus on data protection and risk-based classification of AI. In contrast, the US regulatory design was, until now, sector-specific, while a state-led AI governance framework prevails in China (Michael Abramov, 2024). However, such measures, although helpful in practicing secure coding, as they are currently mainly focused on the security and privacy of the underlying framework and libraries used to train and test AI models, do little to protect against new threats, such as adversarial attacks, deepfake manipulation, and bias in AI-enabled decision-making. Regarding the regulatory shortcomings, the framework recommends the establishment of designated AI regulatory authorities, cybersecurity compliance, and increased cross-border collaboration (Ghiurău & Popescu, 2024).

REFERENCES

Alevizos, L. (2025). Automated cybersecurity compliance and threat response using AI, blockchain and smart contracts. *International Journal of Information Technology, 17*(2), 767–781. https://doi.org/10.1007/s41870-024-02324-9

Almeida, F. (2025). Comparative analysis of EU-based cybersecurity skills frameworks. *Computers & Security, 151*, 104329. https://doi.org/10.1016/j.cose.2025.104329

Baig, A. (2024). *AI Risk Assessment: Strategies, Tools, and Best Practices*. Securiti. https://securiti.ai/ai-risk-assessment/

Bajwa, J., Munir, U., Nori, A., & Williams, B. (2021). Artificial intelligence in healthcare: Transforming the practice of medicine. *Future Healthcare Journal, 8*(2), e188–e194. https://doi.org/10.7861/fhj.2021-0095

Bartsch, S., Milani, V., Adam, M., & Benlian, A. (2024). Algorithmic accountability: what does it mean for AI developers and how does it affect AI development projects. *Hawaii International Conference on System Sciences*. https://doi.org/10.24251/HICSS.2024.702

Belenguer, L. (2022). AI bias: Exploring discriminatory algorithmic decision-making models and the application of possible machine-centric solutions adapted from the pharmaceutical industry. *AI and Ethics, 2*(4), 771–787. https://doi.org/10.1007/s43681-022-00138-8

Binhammad, M., Alqaydi, S., Othman, A., & Abuljadayel, L. H. (2024). The role of AI in cyber security: safeguarding digital identity. *Journal of Information Security, 15*(02), 245–278. https://doi.org/10.4236/jis.2024.152015

Birkstedt, T., Minkkinen, M., Tandon, A., & Mäntymäki, M. (2023). AI governance: themes, knowledge gaps and future agendas. *Internet Research, 33*(7), 133–167. https://doi.org/10.1108/INTR-01-2022-0042

Celi, L. A., Cellini, J., Charpignon, M.-L., Dee, E. C., Dernoncourt, F., Eber, R., Mitchell, W. G., Moukheiber, L., Schirmer, J., Situ, J., Paguio, J., Park, J., Wawira, J. G., Yao, S., & for MIT Critical Data. (2022). Sources of bias in artificial intelligence that perpetuate healthcare disparities—A global review. *PLOS Digital Health, 1*(3), e0000022. https://doi.org/10.1371/journal.pdig.0000022

Christensen, K. K., & Petersen, K. L. (2017). Public–private partnerships on cyber security: a practice of loyalty. *International Affairs, 93*(6), 1435–1452. https://doi.org/10.1093/ia/iix189

Crichton, K., Ji, J., Miller, K., Bansemer, J., Arnold, Z., Batz, D., Choi, M., Decillis, M., Eke, P., Gerstein, D.M., Leblang, A., McGee, M., Rattray, G., & Scott, A. (2024). Securing critical infrastructure in the age of AI. *Center for Security and Emerging Technology.* https://cset.georgetown.edu/publication/securing-critical-infrastructure-in-the-age-of-ai/ https://doi.org/10.51593/20240032

Engler, A. (2023). *The EU and U.S. Diverge on AI Regulation: A Transatlantic Comparison and Steps to Alignment.* Brookings. https://www.brookings.edu/articles/the-eu-and-us-diverge-on-ai-regulation-a-transatlantic-comparison-and-steps-to-alignment/

Folorunso, A., Adewumi, T., Adewa, A., Okonkwo, R., & Nathaniel Olawumi, T. (2024). Impact of AI on cybersecurity and security compliance. *Global Journal of Engineering and Technology Advances*, *21*(1), 167–184. https://doi.org/10.30574/gjeta.2024.21.1.0193

Jariwala, M. (2024). A comparative analysis of the EU AI Act and the Colorado AI act: regulatory approaches to artificial intelligence Governance. *International Journal of Computer Applications*, *186*(38), 23–29. https://doi.org/10.5120/ijca2024923954

Kang, H., Liu, G., Wang, Q., Meng, L., & Liu, J. (2023). Theory and application of zero trust security: A brief survey. *Entropy*, *25*(12), 1595. https://doi.org/10.3390/e25121595

Kaur, R., Gabrijelčič, D., & Klobučar, T. (2023). Artificial intelligence for cybersecurity: Literature review and future research directions. *Information Fusion*, *97*, 101804. https://doi.org/10.1016/j.inffus.2023.101804

Kavanagh, C. (2019). *New Tech, New Threats, and New Governance Challenges: An Opportunity to Craft Smarter Responses?* Carnegie Endowment for International Peace. https://carnegieendowment.org/research/2019/08/new-tech-new-threats-and-new-governance-challenges-an-opportunity-to-craft-smarter-responses?lang=en

Malgieri, G., & Pasquale, F. (2024). Licensing high-risk artificial intelligence: Toward ex ante justification for a disruptive technology. *Computer Law & Security Review*, *52*, 105899. https://doi.org/10.1016/j.clsr.2023.105899

McKinsey. (2023). *Governance and Regulation as Generative AI Advances | McKinsey.* https://www.mckinsey.com/capabilities/risk-and-resilience/our-insights/as-gen-ai-advances-regulators-and-risk-functions-rush-to-keep-pace

Sheehan, M. (2023). *China's AI Regulations and How They Get Made.* Carnegie Endowment for International Peace. https://carnegieendowment.org/research/2023/07/chinas-ai-regulations-and-how-they-get-made?lang=en

5 Securing AI-Integrated Learning Platform

Cyber Threats in the Age of Digital Education

Vijay Krishna Pallaw, Sandeep Kumar Singh, Arun Kumar Singh, Sandeep Saxena, and Divya Mishra

5.1 INTRODUCTION

In the modern era, digital innovations have become essential parts of daily life. They alter the way we seek out data, how we connect to each other, and how we carry on. Consequently, the academic fields have moreover started to alter. Undoubtedly, an increasing number of educational programs are joining the digital environment into their curriculum. Nowadays, many countries are using digital technologies in education. For illustrative purposes, in France, junior classes give a few courses about internet innovations, whereas tenth-standard scholars must clear the test showing their computer skills. Numerous academic services are available for learning management systems such as Moodle [1]. These modifications influenced, apart from various fields, the academic field also. In the past few years, many distinct, intelligent educational innovations have become increasingly popular. Students are using mobiles and tablets rather than notebooks. Faculties are using different learning platforms such as Zoom, Google Classrooms, and Microsoft Teams. Artificial intelligence (AI) has become a highly increasing part of our daily life; undoubtedly, the educational fields are imposingly affected by AU. According to Nwana H. [2], an assessment of how AI can impact individuals and society claims that "Teachers and learners widely utilize AI applications. Nowadays, with a few varieties between school and college settings," involving such devices and advances as teaching robots, smart coaching frameworks, and versatile learning frameworks. Moreover, we will specify such AI applications as versatile expertise building, planning, and carrier instruction [3]. There are numerous definitions of AI that are presently found in the literature. According to Coppin, an AI is a machine that has the capability to adapt to the latest circumstances, deal with developing concerns, resolve issues, answer questions, and perform many other operations that require a certain level of intelligence,

DOI: 10.1201/9781003567585-5

like a human being [4]. According to the author [5], the research and development of Web intelligence and AI focused on distinct components, as well as machine learning and deep learning, to make distributed intelligence and stabilize Web technologies.

5.2 AI IN PRESENT EDUCATION

The announcement of AI comes in the brain, a high-speed computer, a computer with vast processing capacity and versatile manners, like comprehension of sensors and other abilities, which empower it to have humanoid knowledge and behavioral capabilities. It progresses the supercomputer interconnection with a human within the teaching field, which has expanded the implementation of AI. For example, embedded into robots, AI and supporting hardware and software empower robots from the central fundamental unit of education. Undoubtedly, Timms set that the implementation of robots functioning as teachers or co-partner robots are being utilized to educate the young on periodic works, including spelling and altering student capacities [6–8]. Similarly, web-based and online teaching, as counted in distinct educations, has transitioned from essentially accessible content online by scholars to access, read, and complete tasks to clear the examinations. To incorporate intelligent and receptive online-based systems that learns teachers and beginner behavior of change appropriately to improve the educational experience [4, 9–12]. According to M. Chassignol et al. [10], AI in education has been consolidated into management, educating, or learning. The utilization of AI innovations in education is steadily gaining popularity day by day.

5.2.1 Purpose of the Study

With the regular utilization of AI, it is inevitable that it has affected education differently. Considers it looks to survey how the utilization of AI, in its distinct characters, in teaching, has affected or impacted distinctive perspectives of teaching. More significantly, the survey will evaluate how AI can influence teaching, learning, training, and administrative fields of education. It is expected that the study will ensure that AI has cultivated the potential and effectiveness within the execution of managerial jobs in teaching and generally encouraged advanced informational and learning points in education.

5.3 THE ROLE OF AI IN EDUCATION

AI has a remarkable ability to cause significant changes in various distinct fields of the community, with the education field being one that is probably to be highly influenced by AI. AI has been utilized in modern education, especially in management and teaching, impacting student learning. Undoubtedly, AI has been implemented in academic organizations in several ways, including managerial task automation, instructions, and student learning processes. AI has enhanced the functioning and efficiency of administrative works, such as evaluating student tasks, marks, and giving reports on assignments via web-based automation platforms. In other sectors of education, AI has been implemented, including educational program and context development and informational helping innovations—for example, web-based media, robotics, video conferencing, and 3D technologies.

Using online platforms, the learning materials are accessible from any part of the world. The study finding indicates that management, instruction, and learning are more effective and efficient. According to M. Chassignol et al. [10], AI is utilized in academics as a computer and computer-related innovation—for example, the internet. AI in education is basically converting computers into embedded systems, for example robots that perform teacher-like functions. According to Wartman and Combs [13], there is an excellent utilization of AI in medical professions, which requires bringing out scholars to AI by the utilization of technologies in medical education to develop it to achieve the experience in daily life. According to R. C. Sharma et al. [14], AI in academics has been taken as adaptive learning systems and smart education systems, which enhance the characteristics of managerial works, requirements, and learning. According to S. Pokrivcakova [15], another view of AI in instructions, virtual reality (VR), and 3-D technologies is that virtual reality presents vast chances for learning procedure, integrates simulation, and gives a chance to beginners for experimental learning [16]. Table 5.1 describes the various functions of AI in academic scenarios of administration, instruction, and education.

5.3.1 AI in Education Administration

The AI application in academics has several forms and performs distinct operations and imposes a significant effect on the execution of managerial and administrative tasks in education. Smart education systems give a broad range of performances, which empower educators to operate distinct managerial jobs, including student tasks and paper evaluations, grading, and observations. According to R. C. Sharma et al. [14], for online teaching, AI can improve the efficiency and effectiveness of institutional and administrative tasks. According to M. Chassignol et al. [10], AI also provides instructors with distinct managerial responsibilities, as well as plagiarism checking, rating, and grading. AI has improved the efficiency, effectiveness, and performance of distinct managerial tasks, while in the absence of AI, instructors required more time to complete their tasks (Figure 5.1).

TABLE 5.1
Functions of Artificial Intelligence in the Educational Structure

	AI can perform the following roles in modern education
Administration	• Distinguish each student's learning techniques and inclination and support them to build individualized learning plans. • Perform administrative tasks faster, like grading exams and allowing opinions. • Help·teachers in decision-making and data-driven tasks.

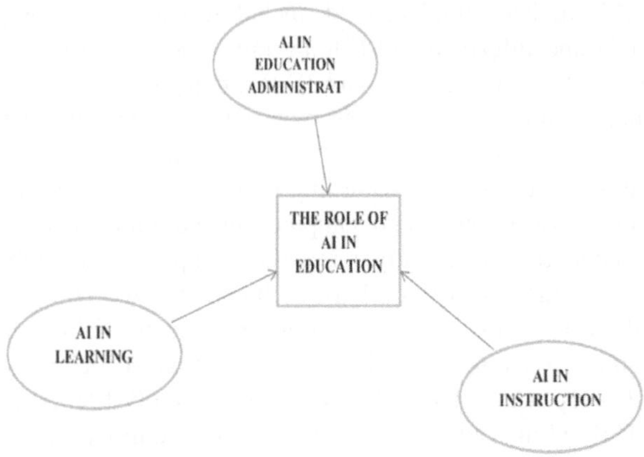

FIGURE 5.1 AI in education administration.

5.3.2 AI in Instruction

The utilization of AI technologies for academic purposes or as an educational mechanism has a crucial effect on educational expression. It has improved the efficiency, strength, and features of the tasks completed by the teacher. AI has cultivated awareness of the quality, efficiency, and effectiveness of teaching. According to S. Pokrivcakova [15], AI has enhanced the educational feature and strength, since modern systems are based on technology, which implies that the substances are decided by the learner's requirements, guaranteeing an optimized learning experience. The evolution and utilization of AI, especially a combination of web-based and online studying platforms, undoubtedly, is driven by awareness of improvement in instructions because AI has given way to improvement and the utilization of superior educational mechanisms for these platforms [17]. According to R. Peredo et al. [12], the same advantages or enhancements to learning are counted in other things that were assessed.

5.3.3 AI in Learning

Learning is a necessary portion of teaching, which is an additional perspective of teaching. After studies and analysis of the different articles, AI has been embraced and executed in student learning and was recognized. Furthermore, programs utilized AI to enhance student learning. A significant way AI has been implemented in enhancing student learning is the personalization of educational programs and substance in compliance with the learner requirements, capabilities, and abilities [17]. Other approaches provide leaners with a more engaging and inclusive experiential learning experience; in this manner, moving forward, learners take up the maintenance of data and the establishment of learning [13, 17]. Another objective of AI in academics has been to remove a few boundaries to access learning via online platforms [14, 17, 18]. There are many distinct mechanisms through which AI

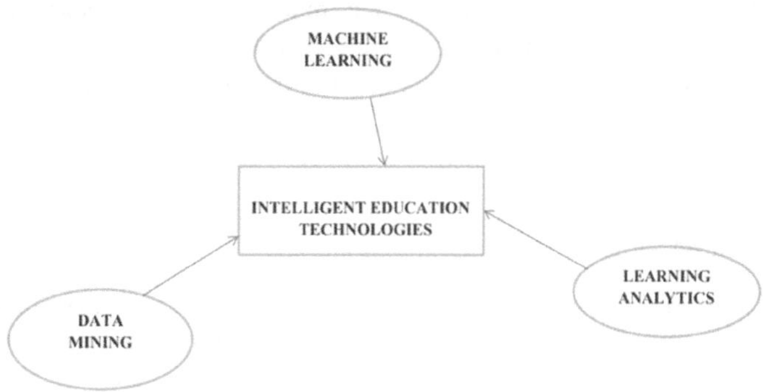

FIGURE 5.2 Intelligent education technologies.

guaranteed the individualization of substance to beginners' potentials and requirements. From the different sources of articles, other AI applications have significant effects on beginners' understandings. As an example, the implementation and utilization of simulation-based learning and smart education systems promote deep learning, enhancing the student's experience. According to Wartman and Combs [13], the utilization of simulation and virtual reality with AI in learning has improved the student's efficiency, performance, and learning experiences.

5.4 INTELLIGENT EDUCATION TECHNOLOGIES

The intelligent innovations for the academic environments are data mining, learning analytics, and machine learning. Two groups have been developed based on educational data mining and learning analytics (Figure 5.2) [19].

5.4.1 Machine Learning

A fundamental of machine learning is knowledge disclosure, parsing based on training data, and creating significant patterns and organized knowledge. For example, machine learning may help to suggest students. It credits students' accomplishments, desires, and performances to coordinate institutions where they can be highly developed. Specifically, for student evaluation, image recognition and machine learning prediction may be utilized to review student tasks and exams more quickly and with greater accuracy than a person.

5.4.2 Learning Analytics

Learning analytics concentrates on the data from the features of scholars and the information purpose from the learning demonstrate and information field model. The learning analytics concept presents modern innovation. The reason is to tailor an academic strategy to the person learners require and capacities, such as mediating with scholars by chance or giving input and guidelines substance [20, 21]. It utilizes the procedure associated with data science, machine learning, learning sciences, and data visualizations. In addition to competence-based learning, learning analytics also employs the flexible capabilities of AI to memorize. Learning analytics will be progressively utilized and combined with advanced procedures to sustain learning for scholars, educators, organizations, and colleges.

5.4.3 Data Mining

Data mining seeks to create taxonomical and automatic feedback for beginners. AI-based instructional data mining, motives for creating intrinsic association laws, and giving comprehension objects to the students to meet their individual requirements. For example, a student's demographic characteristics information and grading information may be examined from a couple of printed assignments [22]. Besides, data mining is an effective tool to advance the learning process and information dominance, leading to an exemplary realization of the academic circumstances. In other words, data mining may be considered as pattern recognition and predictive modeling applied to extract hidden knowledge, which enables teachers to make adjustments that enhance curriculum improvement within the academic system.

5.5 IMPACT OF AI IN EDUCATION

AI is now an essential part of daily life. This innovation encompasses us. The implementation of AI in modern education systems can help students build their skills and acquire more knowledge in various subjects. AI has changed the students' way of learning, as they no longer need to be physically present in the classroom since they have access to learning material through the internet [23]. As already specified, AI permits teachers to spend additional time with students by taking over some administrative work (Figure 5.3).

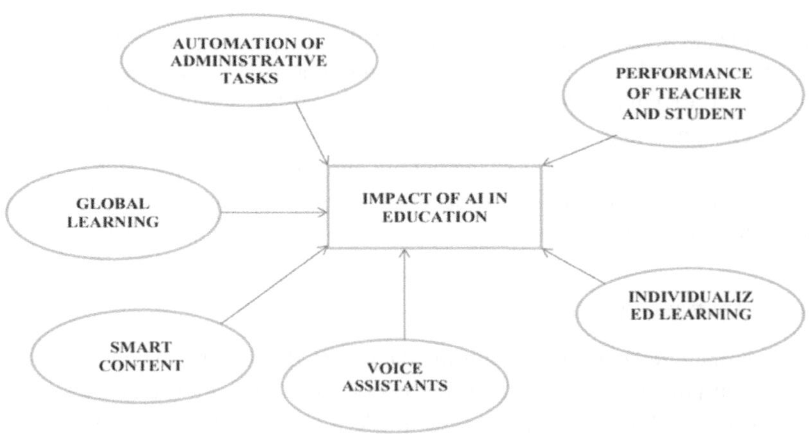

FIGURE 5.3 Impact of AI in education.

5.5.1 Automation of Administrative Tasks

AI has a remarkable ability to automate the undertaking of managerial responsibilities for both educators and organizations. Teachers devote additional time to grading exams, evaluating assignments, and offering essential feedback to their students. But AI-based technologies may be utilized to automate the grading tasks where different exams are involved. It implies that teachers would have additional time with their students instead of investing more time grading them. Software developers are creating the best method to grade written responses and essays. A smart tutoring system offers a broad range of performances that empower teachers to achieve distinct managerial tasks, including grading and giving feedback [24]. AI also allows instructors to perform distinct managerial activities, including plagiarism checking, reviewing, and evaluating. AI has remarkably decreased the paperwork and workload on teachers, especially in several managerial tasks.

5.5.2 Performance of Teacher and Student

As a smart system, it would be interesting to see how AI will influence the functioning of teachers and students. AI systems will work well in facilitating the load on teachers. AI-based technologies support teachers in analyzing the syllabus and course materials to offer customized content [10]. After analyzing, AI systems can produce and grade exams. This would inevitably free instructors to concentrate on scholarly achievements. In individual learning and independent learning, AI determinations may superiorly examine learning data in supporting teachers to form customized learning schemes for every scholar. Most scholars have been handled correspondingly by the conventional teaching system [22]. AI would support finding the custom-made educational strategy for every scholar based on their identities, qualities, and interrelated skills. Hence, every scholar can enhance their learning performance. Whereas expanding knowledge of students, it moreover supports the growth of knowledge systems by improving learning capacity, creativity, and propensities. Considering the independent capability and career path, students can get higher marks and acquire skills that are appropriate within the real world. Additionally, in the learning process, AI enables education to access student understanding based on the entire environment of learning tools [25].

5.5.3 Individualized Learning

AI permits concentrating on the personal requirements of the students. Many of the education platforms, such as Carnegie Learning, contribute to AI to supply more individualized courses. Nowadays, it is conceivable to make personal instructions, testing, and criticism. Thus, beginners operate with data they are prepared for and cover disruption in their intelligence. Therefore, as AI becomes more intelligent, it can be conceivable to check and examine student facial gestures. Voice assistants, such as Google Home, Apple Siri, and Amazon Alexa, allow interaction with different learning materials without communication with a faculty. Therefore, it is conceivable to utilize a teaching platform anytime and anywhere. Furthermore, using such assistants, it is very interesting and inspiring for beginners, so they are involved in the teaching procedure. The term "innovative content" stands for different study materials, from the digitalized course reading to customize interfacing. We describe two examples.

(a) Context Technologies, Inc. is a development firm that collaborates with artificial intelligence. Its primary objectives are to enhance the user experience and automate trade forms.
(b) Netex Learning is another firm that concentrated on making innovative content platforms. NETEX learning provides customized cloud platforms with conferences, virtual training, and others.

5.6 EDUCATING STRATEGIES: FROM PERSONALIZED LEARNING TO EDUCATIONAL ROBOTS

There is a tremendous challenge in teaching students, in that everybody has a distinct learning and understanding of instructions. Therefore, earlier in front of substance may effectively be bored, whereas others could be misplaced in modern terms and circumstances. AI gives an opportunity to keep away from this difficulty called personalized learning [26, 27]. Personalized learning is a technique that tailors academic substance to the individual requirements of a particular scholar [28] and facilitates educational significance for every student's need, the stamina of learning, and knowledge level. Mainly, gamification components such as leaderboards and points may support the report of student improvement and resolve the issue of adjusting student speed of understanding modern educational materials (Table 5.2).

TABLE 5.2
A Few Cases of Personalized Learning Platforms and Tools

S. No.	Tool Titles	Objectives	Target Consultation
1	Knewton [29]	The analysis-based platform creates suggestions for scholars directly on their studying fashion and feeds substance to their requirements	College and higher education students
2	Highlights [30]	Cloud concept-based learning platform that tracks scholar's growth over different substance suppliers	College and higher education students
3	Waston Education Classroom [31]	Cloud service solution facilitates professors to impart personalized learning to enhance students' results by picking up information to realize scholars' requirements and customize learning exercises	College and higher education students
4	Cerego [32]	An algorithm-based platform that evaluates individual memory maintenance and makes an ideal plan for effectively guided learning	Companies and universities

TABLE 5.3
Examples of Educational Robots

S.No.	Robot Titles	Objectives	Target Consultation
1	Root [33]	The main objective of the Root robot is to help students learn how to code.	Students at distinct school levels
2	Cubelets [34]	Cubelets are robots that train students to plan and think logically as they build a robot.	Students as young as four
3	Ozobot Bit [35]	Ozobot Bit allows students to play while learning coding and logic skills.	School students
4	mBot [36]	mBot is a robot kit supporting students in distinct aspects of robotics like electronics, assembly, and programming.	Students at distinct school levels
5	NAO [37]	NAO is a humanoid robot that can observe, listen, and teach any subject.	Students from age 5 up to university level

All other feedback fundamentally revolves around the entire thought that a deeply inspired scholar does not even require a customized learning system because this scholar finds every essential data needed for his learning. The fame of the customized learning these days progresses not from students asking but from the software company that needs to trade software. The academic sector is one of the areas that implemented robots for the functioning operations [19, 21]. Table 5.3 describes many illustrations of robots that have been utilized to educate students of different ages. The cause of their fame could be effectively clarified by student-companion interface, availability, and attractiveness. Robots can educate students in programming, logic, reasoning, problem-solving, and robotics. It is accomplished through experience, and some of the time by utilizing a few of the latest techniques. The same feedback allowing human connection may be enforced; the AI robot substitution for an instructor cannot result in students' understanding of social mechanisms. Every scholar has distinct learning capacities. The best instructor must make an effort to provide data in a way that reaches each scholar. Some scholars may have behavioral or intellectual disease that prevents or make complex these processes.

5.7 CYBER THREATS IN EDUCATION: CHALLENGES AND SOLUTIONS

The education sector, as a decisive sector attacked by cyber threats, is selected as the main focus of this research because it has become critically important in the age of digital transformations. Schools, colleges, and universities today have come to depend on online learning platforms, cloud solutions, and digital communication in their teaching, administrative, and learning processes. Despite these developments, they have brought about positive impacts in delivering education services by improving their access and the efficiency of delivery. They have also uncovered institutions to the following cybersecurity threats: data losses, ransomware, phishing scams, and distributed denial of Service (DDoS) attacks.

5.7.1 Data Breaches and Privacy

Cyber threats that are very common in the educational institutions include data breaches. Schools, colleges, and other educational organization process and store significant amounts of personal information about students, financial information, and property and research outcomes. Intravenous breakdown of such systems results in data loss; this may be PII or sensitive research information. One of the most recent ransomware attacks, which at least took place in 2020, involved UC station University of California, San Francisco. The university had to be dismissed and pay $1: 14M, for the release of its data) [38]. They do not only lead to monetary loss but also erode the reputation of the institution and undermine student privacy and data protection [39].

5.7.2 Ransomware Attacks

Recently, in the education environment, cyber attackers have used ransomware to encrypt institutional data and claimed that the data could only be decrypted if the attackers were paid a sum of money. For example, the University of Utah suffered the cyberattack in 2020 and decided to pay a ransom of $457,000 to the enemy to stop the revelation of student/staff records [38]. Such operations can directly affect academic business and learning, stall research activities, and result in serious financial and reputational loss to the institutions [39].

5.7.3 Phishing Scams

Phishing attacks are among the most common cybersecurity threats faced by educational institutions. Hackers utilize deceptive emails to trick students, faculty, and staff into disclosing secret data, such as financial data or login credentials. Phishing scams have significantly increased during the COVID-19 pandemic, targeting the growing number of online learning platforms and communication systems [40]. Without adequate cybersecurity awareness, many users fall victim to these attacks, which often lead to unauthorized access to institutional systems and personal accounts. Another prevalent cyber threat is Distributed Denial of Service (DDoS) attacks, which overwhelm institutional networks with traffic, making services such as online learning platforms and administrative systems inaccessible. The U.S. saw a sharp increase in DDoS attacks on K-12 schools during the pandemic, as many institutions shifted to remote learning [41]. These attacks can disrupt not only educational services but also essential administrative functions, such as grading and communication with parents and students. The insider threats are equally as important but more often not talked about. Unprivileged users may include employees or students who have access to vital systems and organizations' databases, and they take advantage of this access to tamper with capital resources or help extraneous attackers. According to Schneier [42], among the key areas that make educational institutions so susceptible to cyber threats, the absence of cybersecurity training for staff, students, or other administrators is significant. Anderson and Rainie [43] found that more than 90% of K-12 school administrators had not been trained enough in cybersecurity; students consequently fall prey to phishing scams, password theft, and malware. Speaking of the solutions, starting with the cybersecurity training should be named as one of the priorities because, following this training, individuals can distinguish simple threats, for example, emails that seem rather suspicious or using obvious passwords.

5.8 CONCLUSION

The main aim or motive of this work is to evaluate the effect of AI on education. It also describes how AI can assist in resolving the problems of students and realizing how to help them enhance the interest of a collectivity and plan modern academic experiences. The growth and utilization of computers and computer-based innovations harbinger research and development leading to the development and utilization of AI technologies in various fields.

AI has been broadly embraced and utilized in academic sectors, especially in educational organizations, which were concentrated on calculating the effect of AI on managerial tasks and learning viewpoints of instruction. The applications of AI technologies are providing numerous chances for developing online courses. Initially, AI in education was applied through computers and computer-based systems and later through web-based teaching platforms. Embedded systems have made it feasible to utilize robots in collaborative robots as a faculty, including chatbots to carry out the functions of a teacher. Using these platforms and tools has improved the faculty's success, efficiency, and performance and improved instructional quality. Since AI has permitted the personalization of learning material, it has enhanced students' learning experiences. An individualized learning environment enhances the characteristic procedure, but it also provides a chance for scholars with problems like dyslexia or other learning disabilities to think about highly successful. It also supports academic substance to scholars' requirements, creating more personalized study plans. Overall, it confirms that AI technologies and mechanisms have been applied in various parts of academic procedures, including substance evolution, teaching strategies, scholar evaluation, and communication between students and teachers. Finally, AI has significantly impacted education, especially in the areas of management, guidance, and learning across the educational sector.

REFERENCES

1. Moodle Statistics [Electronic resource]. URL: https://moodle.net/stats/.
2. H. Nwana, Intelligent tutoring systems: an overview. *Artificial Intelligence Review*, vol. 4, pp. 251–277, 1990.

3. I. Arroyo, D. G. Cooper, W. Burleson, B. P. Woolf, K. Muldner, and R. Christopherson, "Emotion sensors go to school," *Artificial Intelligence in Education*, vol. 200, pp. 17–24, 2009.
4. B. Coppin, *Artificial Intelligence Illuminated.* Boston, MA, USA: Jones and Bartlett, 2004.
5. V. Devedžic, "Web intelligence and artificial intelligence in education," *Educational Technology & Society*, vol. 7, no. 4, pp. 29–39, 2004.
6. M. J. Timms, "Letting artificial intelligence in education out of the box: Educational cobots and smart classrooms," *International Journal of Artificial Intelligence in Education*, vol. 26, no. 2, pp. 701–712, Jan. 2016.
7. H. Snyder, "Literature review as a research methodology: An overview and guidelines," *Journal of Business Research*, vol. 104, pp. 333–339, Nov. 2019.
8. Y. Fang, P. Chen, G. Cai, F. C. M. Lau, S. C. Liew, and G. Han, "Outagelimit-approaching channel coding for future wireless communications: Root-protograph low-density parity-check codes," *IEEE Vehicular Technology Magazine*, vol. 14, no. 2, pp. 85–93, Jun. 2019.
9. D. Wang, H. Han, Z. Zhan, J. Xu, Q. Liu, and G. Ren, "A problem solving oriented intelligent tutoring system to improve students' acquisition of basic computer skills," *Computers & Education*, vol. 81, pp. 102–112, 2015.
10. M. Chassignol, A. Khoroshavin, A. Klimova, and A. Bilyatdinova, "Artificial intelligence trends in education: A narrative overview," *Procedia Computer Science*, vol. 136, pp. 16–24, Jan. 2018.
11. H. T. Kahraman, S. Sagiroglu, and I. Colak, "Development of adaptive and intelligent Web-based educational systems," in *Proceedings of the 4th International Conference on Applied Information Communication Technology*, pp. 1–5, Oct. 2010.
12. R. Peredo, A. Canales, A. Menchaca, and I. Peredo, "Intelligent Webbased education system for adaptive learning," *Expert Systems with Applications*, vol. 38, no. 12, pp. 14690–14702, Nov. 2011.
13. S. A. Wartman and C. D. Combs, "Medical education must move from the information age to the age of artificial intelligence," *Academic Medicine*, vol. 93, no. 8, pp. 1107–1109, Aug. 2018.
14. R. C. Sharma, P. Kawachi, and A Bozkurt, "The landscape of artificial intelligence in open, online and distance education: Promises and concerns," *Asian Journal of Distance Education*, vol. 14, no. 2, pp. 1–2, 2019.
15. S. Pokrivcakova, "Preparing teachers for the application of AI-powered technologies in foreign language education," *Journal of Language and Cultural Education*, vol. 7, no. 3, pp. 135–153, Dec. 2019.
16. United Nations Education Scientific and Cultural Organization (UNESCO). *How Can Artificial Intelligence Enhance Education?* [Online]. Available: https://en.unesco.org/news/how-can-artificialintelligence-enhance- education (2019).
17. T. A. Mikropoulos and A. Natsis, "Educational virtual environments: A ten-year review of empirical research (1999–2009)," *Computers & Education*, vol. 56, no. 3, pp. 769–780, Apr. 2011.
18. S. Serholt, C. A. Basedow, W. Barendregt, and M. Obaid, "Comparing a humanoid tutor to a human tutor delivering an instructional task to children," in *Proceedings of the IEEE-RAS International Conference on Humanoid Robots*, pp. 1134–1141, Nov. 2014.
19. J. Estevez, G. Garate, and M. Graña, "Gentle introduction to artificial intelligence for high-school students using scratch," *IEEE Access*, vol. 7, pp. 179027–179036, 2019.

20. Y. Kim, T. Soyata, and R. F. Behnagh, "Towards emotionally aware AI smart classroom: Current issues and directions for engineering and education," *IEEE Access*, vol. 6, pp. 5308–5331, 2018.
21. T. Yi-Shan and D. Gasevic, Learning analytics in higher education— Challenges and policies: A review of eight learning analytics policies," in *Proceedings of the Seventh International Learning Analytics & Knowledge Conference*, pp. 233–242, Mar. 2017.
22. *Global Development of AI-Based Education, Deloitte Research*, Deloitte China: Deloitte Company, 2019.
23. M. M. L. Cairns "Computers in education: The impact on schools and classrooms," in *Life Schools Classrooms*. Singapore: Springer, pp. 603–617, 2017.
24. V. Rus, S. D'Mello, X. Hu, and A. Graesser, "Recent advances in conversational intelligent tutoring systems," *AI Magazine*, vol. 34, no. 3, pp. 42–54, Sep. 2013.
25. K. Flamm, *Creating the Computer: Government, Industry, and High Technology*. Washington, DC, USA: Brookings Institution Press, 1988.
26. Newton C. Can AI fix education? We asked Bill Gates // The Verge, 2016.
27. M. King, R. Cave, M. Foden, M. Stent, et al., Personalised education: From curriculum to career with cognitive systems, IBM Education, UK, 2016.
28. D. Hutchins, "AI Boosts Personalized Learning in Higher Education." *Educational Technology*, 2017.
29. Knewton [Electronic resource]. URL: https://www.knewton.com/.
30. Highlight: Our Personalized Learning Platform [Electronic resource]. URL: https://www.edelements.com/blog/highlight-our-personalized-learning-platform.
31. Watson Education Classroom [Electronic resource]. URL: https://www.cerego.com/.
32. Cerego: Intelligent Learning Software for Companies & Educators [Electronic resource]. URL: https://www.cerego.com/.
33. Root Robotics I The Robot to Get Anyone Coding [Electronic resource]. URL: http://www.codewithroot.com/.
34. Cubelets Robot Construction for Kids I Modular Robotics [Electronic resource]. URL: https://www.modrobotics.com/cubelets/.
35. Ozobot I Robots to Code, Create, and Connect [Electronic resource]. URL: https://ozobot.com/.
36. Educational Robot Kit: mBot I Makeblock–Let Creation Be a Way of Life [Electronic resource]. URL: https://www.makeblock.com/steam-kits/mbot.
37. B. I. Edwards and A. D. Cheok, "Why not robot teachers: Artificial intelligence for addressing teacher shortage," *Applied Artificial Intelligence*, vol. 32, no. 4, pp. 345–360, 2018.
38. K. Zetter, "University of Utah pays $457,000 to prevent ransomware data leak," *Wired*, 2020. [Online]. Available: https://www.wired.com. Accessed: Sep. 23, 2024.
39. R. E. Campos and J. P. Leal, "Cybersecurity challenges in higher education: Data breaches and ransomware," *Computers & Security*, vol. 102, p. 102171, 2021.
40. R. N. Katz, "Phishing in higher education: A case study on prevention and awareness," *Journal of Cybersecurity Education*, vol. 5, no. 2, pp. 34–42, 2021.
41. FBI and CISA, "Joint cybersecurity advisory: Surge in DDoS attacks targeting K-12 schools," *Cybersecurity & Infrastructure Security Agency*, 2020. [Online]. Available: https://www.cisa.gov. Accessed: Sep. 23, 2024.
42. B. Schneier, "The insider threat in education: Mitigating the risk," *IEEE Security & Privacy*, vol. 18, no. 5, pp. 74–78, 2020.
43. C. Anderson and L. Rainie, "The lack of cybersecurity preparedness in K-12 schools," *Pew Research Center*, 2020. [Online]. Available: https://www.pewresearch.org. Accessed: Sep. 23, 2024.

6 Blockchain-Decentralized Approach for Crowdfunding Using ECDSA with P-256 and SHA-256 Algorithm

V. Subbulakshmi, M. Parvathy, M. Aanandalakshmi, and R. Neharika

6.1 INTRODUCTION

Crowdfunding has become a popular alternative to traditional financing, enabling creators and organizations to raise capital from a global pool of contributors. Platforms [1–4] such as Kickstarter, GoFundMe, and Indiegogo have demonstrated the model's potential by democratizing access to funding. However, these systems rely heavily on centralized intermediaries, which introduces significant limitations.

Centralized crowdfunding platforms pose challenges including opaque fund management, high transaction fees (typically 4–8%), delayed withdrawals, and exposure to fraud. Users must trust the platform to act honestly, yet there is no cryptographic assurance that funds will be [5] used as promised. High-profile instances of fund misappropriation and platform shutdowns have exposed the fragility of this trust model.

Blockchain-based crowdfunding aims to resolve these issues by replacing intermediaries with self-executing smart contracts, offering tamper-resistant records [6] and programmable fund release conditions. Projects such as KickICO, Gitcoin, and Juicebox have made strides in this space, but gaps remain. These include the following:

- Lack of FIPS-validated authentication mechanisms.
- Inadequate support for tiered or conditional disbursements.
- Absence of hash-linked metadata for auditable campaign tracking.
- Limited testing on actual blockchain networks with usable interfaces.

DOI: 10.1201/9781003567585-6

To address these challenges, we present EtherRaise, a decentralized crowdfunding platform designed with a security-first approach. EtherRaise leverages:

- ECDSA-P-256 for strong, NIST-approved digital signatures.
- SHA-256 for tamper-proof data verification.
- Smart contracts deployed on the Sepolia testnet for programmable fund control.
- A Next.js and Thirdweb front end that eliminates custodial wallets and simplifies user interaction.

Our contributions include the following:

(i) A dual-layer cryptographic scheme (ECDSA + SHA-256) to secure transactions and metadata.
(ii) A gas-optimized Solidity contract that processes pledges under 45k gas.
(iii) A non-custodial UI that enables wallet-based sign-in and signing.
(iv) A threat analysis covering signature forgery, replay attacks, and DoS vectors.
(v) A comparison with existing platforms showing improved latency and cost-efficiency.

EtherRaise demonstrates how integrating secure cryptographic primitives with decentralized infrastructure can produce a crowdfunding platform that is transparent, auditable, and resilient.

6.2 RELATED WORKS

Blockchain has been increasingly adopted in crowdfunding to improve transparency and reduce dependency on centralized intermediaries [7]. Early platforms such as KickICO and Gitcoin leverage Ethereum smart contracts to automate fund management, offering a level of decentralization and trustless interaction. However, most of these platforms lack strong cryptographic user authentication and secure transaction verification mechanisms.

Blockchain applications in a variety of fields, including digital loans, mobile crowdsourcing, land registration, and blood donation systems, have been covered in several research studies [1–5]. These initiatives place a strong emphasis on traceability and transparency, but they ignore the use of cutting-edge cryptographic technologies such as Secure Hash Algorithm-256 (SHA-256) and Elliptic Curve Digital Signature Algorithm (ECDSA), which are essential for ensuring transaction security and non-alteration in sensitive and commercial products such as crowdfunding.

The blockchain system's performance and consensus algorithms are improved in the works of [6, 8, 9]. These works examine the alternatives of proof-of-work/proof-stake and the generation of hash keys with time efficiency. Despite their efforts to improve scalability and efficiency, these projects do not currently offer techniques for safe user verification. Furthermore, decentralized access using blockchain-enabled digital wallets such as MetaMask does not execute signature verification within smart contract logic, hence facilitating spoofing and unauthorized access.

Researchers have also proposed decentralized identity models and PKI infrastructures [10, 11] to ensure data integrity in other domains. However, these approaches are rarely extended to crowdfunding platforms. Without user-level cryptographic validation, platforms remain vulnerable to misuse, such as unauthorized fund access or forged campaign records.

Furthermore, domain-specific crowdfunding studies [12–17] (e.g., for startups, charity projects, and educational loans) focus on blockchain's transparency but fail to implement or evaluate digital signature-based authentication. These platforms often rely on frontend-level identity management, which can be easily compromised.

In summary, while prior research validates blockchain's value in enhancing trust and automation in crowdfunding, the integration of strong, decentralized cryptographic techniques, especially ECC-based digital signatures and hash functions for transaction verification, is underexplored. This motivates the development of EtherRaise, a platform that bridges this gap by embedding ECDSA (P-256) and SHA-256 directly into smart contracts and user workflows.

6.3 PROPOSED WORK

The proposed system, EtherRaise, is a decentralized crowdfunding platform designed to enhance transparency, reduce platform dependency, and secure financial transactions using blockchain and advanced cryptographic techniques. By eliminating intermediaries and deploying self-executing smart contracts on the Ethereum blockchain, EtherRaise ensures that campaign creation, contributions, and disbursements are trust-minimized and verifiable [18].

- The architecture of EtherRaise is shown in Figure 6.1, where two smart contracts, the crowdfunding factory contract and the individual campaign contract, are deployed on the Sepolia Test Network (chain ID 11155111). This testnet is chosen for its fast block confirmation time of approximately 12 seconds, which facilitates responsive interaction during experimentation and prototyping [19]. The smart contracts are written in Solidity and optimized to execute primary functions, such as campaign creation and pledge handling, within a gas cost of under 45,000 units.

6.3.1 Enhancing Transaction Security with ECDSA

The ECDSA is a cornerstone of EtherRaise's authentication framework. Utilizing the NIST-approved P-256 curve, ECDSA provides 128-bit security with compact 64-byte signature sizes, making it suitable for blockchain applications where performance and signature verification are critical. In EtherRaise, all transactions—whether initiated by contributors or campaign creators—are signed using a private key and validated using the corresponding public key [20]. This signature ensures that only authorized [21] users can initiate operations on the smart contracts, thereby

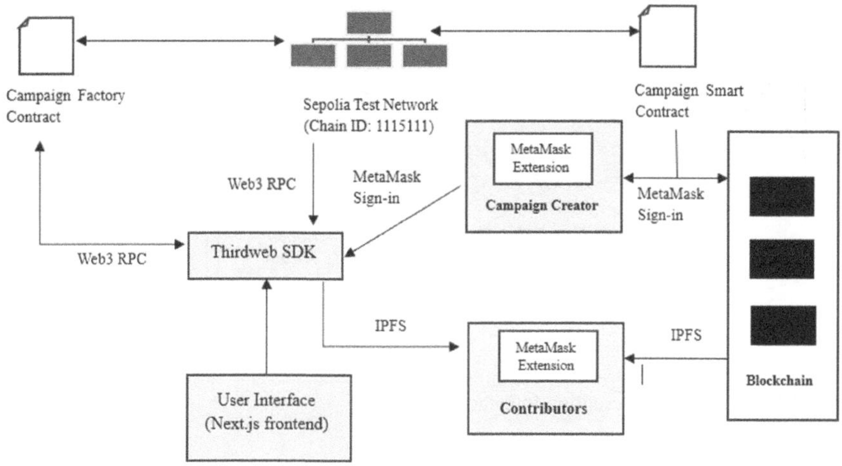

FIGURE 6.1 EtherRaise component stack.

eliminating the risk of unauthorized fund transfers. EtherRaise integrates the ECDSA library from OpenZeppelin v5.0 (ECDSA.sol) to achieve secure wallet-based interactions and to prevent forgery or impersonation in all pledge-related activities.

6.3.2 Contract Creation Interface

EtherRaise ensures transaction and campaign integrity using the SHA-256 cryptographic hash function, implemented in both the frontend (via Node.js crypto) and Solidity (keccak256). Each critical action—such as createCampaign, editTiers, or togglePause—is hashed before being processed by the smart contract. This ensures immutability and auditability of user interactions, even under adversarial conditions.

As illustrated in Figure 6.2, the interface [22] permits creators to dynamically manage campaign states. Every transaction payload (e.g., campaign metadata) is hashed client-side before contract submission, providing a verifiable digital fingerprint on-chain. This mechanism supports Ethereum's trust model by preventing tampering and allowing transparent verification of all user-submitted data.

On test hardware (Apple M2, Node.js v20), SHA-256 throughput reached ~330 MB/s, validating the feasibility of near real-time hashing for transaction volumes encountered in crowdfunding use cases.

6.3.3 SHA-256 Pseudocode for Crowdfunding Transactions

The SHA-256 algorithm operates on structured input data, such as the donor address, campaign ID, and donation amount. This input is padded, split into 512-bit chunks, and processed through a series of bitwise and logical functions to produce a 256-bit hash. The pseudocode outlines initialization steps, message preprocessing, and

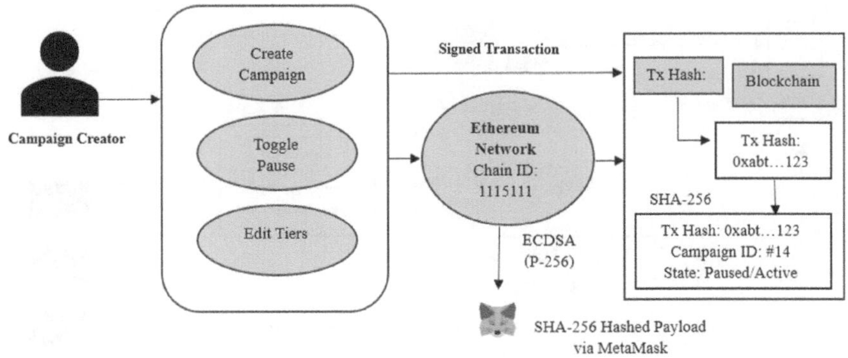

FIGURE 6.2 Contract creation interface.

compression loops, following the standard as defined in FIPS PUB 180-4. While technical in nature, this implementation provides the foundation for securely hashing user interactions within the EtherRaise ecosystem. On a mid-range processor such as Apple M2, the SHA-256 hash throughput is measured at approximately 330 MB/s, ensuring low computational latency during real-time fundraising events.

6.3.4 Contributions via the Decentralized Interface

EtherRaise's decentralized interface enables contributors to interact directly with the Ethereum network using their MetaMask wallets. Built using Next.js and integrated with the Thirdweb SDK, the interface displays live campaign data such as funding goals, time remaining, contribution progress, and reward tiers. Contributors can select campaigns, choose pledge levels, and sign transactions using their wallets, eliminating the need for custodial intermediaries. Once confirmed, each transaction is recorded immutably on-chain and verified via ECDSA and SHA-256 mechanisms. The platform's UI provides real-time feedback, including transaction hashes and confirmations, offering users transparency and traceability throughout the crowdfunding lifecycle.

6.4 RESULTS AND DISCUSSION

To evaluate the performance and security of EtherRaise, we conducted experiments on the Sepolia Ethereum test network. The testing environment included 28 active crowdfunding campaigns and over 400 individual contributions made through MetaMask wallets under both Wi-Fi and 4G network conditions. Key performance metrics, including transaction latency, gas consumption, and cryptographic integrity, were recorded and analyzed.

TABLE 6.1
Gas Consumption (in Gas Units)

Function	With Sign Verification Batching	Without Sign Verification Batching
Campaign creation	43,812	52,015
Contribution	44,932	50,100
Withdrawal	39,151	42,750

6.4.1 Performance Metrics

The system demonstrated strong performance in terms of cost efficiency and speed. Table 6.1 summarizes the key operational metrics captured during campaign creation, contribution, and withdrawal functions:

> These results confirm the impact of ECDSA signature batching on gas optimization, contributing to reduced operational costs for users.

Additional measured metrics include:

- MetaMask Confirmation Time (under 4G): 2.6–3.1 seconds
- Zero-trust Disbursement Latency: < 12 seconds (from pledge to contract confirmation)
- Average Pledge Amount: 0.14 ETH
- Median Gas Fee per Contribution: 0.00082 ETH (~₹145)
- Transaction Success Rate: 100% (no failed contributions)
- Security Incidents Prevented: 3 replay attack attempts (blocked by nonce verification)

6.4.2 Usability and Transparency

The platform's front-end, built using Next.js and connected via the Thirdweb SDK, enabled smooth interaction between contributors and the blockchain. A total of 20 participants unfamiliar with blockchain tools were asked to interact with EtherRaise; 85% successfully completed contributions without external help. The primary friction point reported was MetaMask wallet onboarding, which is expected and can be mitigated with better UX walkthroughs.

Access control was enforced such that campaign details remained inaccessible until authentication, ensuring a secure and personalized experience. All transaction data—sender address, amount, campaign ID, and status—was permanently stored on-chain and accessible via the dashboard for auditing.

6.4.3 Security and Cryptographic Verification

Security is ensured through the use of ECDSA with the P-256 curve (OpenZeppelin 5.0 ECDSA.sol), providing 128-bit equivalent security with a 64-byte signature. All transactions are signed by the sender and verified on-chain using public key recovery. Signature verification failures were nonexistent in the evaluated period, confirming proper wallet and key handling.

Additionally, SHA-256 is used to hash all sensitive campaign metadata, providing tamper-resistance and enabling integrity checks. The system correctly rejected three attempted replay attacks during testing, verifying the effectiveness of nonce-based protection mechanisms.

6.4.4 Comparative Analysis

Compared to existing platforms, EtherRaise offers substantial improvements (Table 6.2):

These findings demonstrate that EtherRaise reduces operational cost by over 60% compared to centralized platforms while improving transparency and cryptographic security.

6.4.5 Relevance to Predictive Electronics

EtherRaise's architecture is designed for low-overhead environments. SHA-256 hashing throughput, measured at ~330 MB/s on a mid-tier Apple M2 node, suggests compatibility with embedded nodes. Future deployment will evaluate the use of secure elements such as Infineon OPTIGA TPM and Ledger Secure Enclaves for offline signing and campaign verification, enabling use cases in predictive electronics systems that require cryptographically verifiable, decentralized funding.

6.5 CONCLUSION

This chapter presented EtherRaise, a decentralized crowdfunding platform designed to overcome the limitations of traditional centralized systems such as high transaction fees, escrow opacity, and fraud exposure. The system leverages a dual-layer cryptographic architecture, combining ECDSA-P-256 for secure authentication and

TABLE 6.2
Platform Comparison

Platform	Avg. Fee	Transparency	Custodial Wallets	Audit-Friendly?
Kickstarter	5–8%	Low	Yes	No
Gitcoin	2–4%	Medium	Optional	Partial
EtherRaise	**<2%**	**High**	**No**	**Yes**

SHA-256 for tamper-proof transaction state hashing. Smart contracts deployed on the Sepolia testnet (Chain ID: 11155111) automate fund management and disbursement, eliminating the need for intermediaries.

Experimental results demonstrate EtherRaise's capability to reduce transaction fees by over 60% compared to centralized platforms, while enabling trustless disbursements with sub-12 second latency. Signature verification, nonce-based replay protection, and immutable data storage together ensure strong security guarantees. In contrast with prior blockchain-based crowdfunding attempts, EtherRaise provides formal cryptographic validation, gas-optimized contract logic (<45k gas per pledge), and a non-custodial front-end via MetaMask and Thirdweb SDK integration.

By demonstrating secure, verifiable crowdfunding workflows, EtherRaise lays the groundwork for decentralized, fraud-resistant funding mechanisms suitable for both consumer and embedded-device contexts.

6.6 FUTURE WORK

Through a series of fundamental enhancements, the subsequent phase of the development of a decentralized crowdfunding platform would be to prioritize its scalability, security, and user experience. Initial advancements will involve the use of Layer-2 scaling solutions such as Optimistic Rollups and zk-Rollups to reduce costs and expedite transaction processing. By employing these technologies, the platform will enhance transaction processing speed and reduce costs while maintaining decentralization.

The security system will undergo a significant enhancement, as the project intends to include decentralized identity verification (DID) technology for safe private authentication. One additional layer of defense against malicious activity is provided by this system, which enables users to verify their identities without disclosing any additional information. Post-quantum cryptography techniques will be examined to maintain the platform's security against emerging cybersecurity threats.

This platform will implement a milestone-based money distribution strategy to increase openness and confidence in the pledge process. Under this model, campaign funds are delivered progressively depending on pre-established objectives. Consequently, the project developer will be compelled to utilize the amassed funds more judiciously, hence diminishing the likelihood of further fraud. A refund method will be instituted that operates automatically, allowing contributors to recover their payments if a campaign fails to meet its objective, so enhancing credibility. Decentralized management will be community-led, enabling on-chain voting for users to cast votes on choices. This alteration will restore authority to the community, enabling them to propose and vote on the platform's direction, necessary policy amendments, and allocation of financing. This governance paradigm will markedly democratize and enhance transparency within the crowdfunding community.

AI-powered advanced analytics will be used to offer a more detailed look at funding trends, campaign performance, and donor behavior. This analysis will enable the campaign creator to refine their fundraising strategy and provide more pertinent project recommendations to contributors. Employing AI-driven fraud detection technologies to identify suspicious conduct and prohibit malicious users from accessing the service.

A mobile application will be developed to improve accessibility, allowing anyone to effortlessly create, browse, and participate to campaigns on their mobile devices. The program will offer an intuitive UI and real-time notifications, coupled with digital wallets for swift transactions. Integration with historical banking infrastructure will enable easy conversion between cash and cryptocurrency transactions. This tool will allow anyone lacking familiarity with cryptocurrencies to engage in blockchain crowdfunding.

Elements of gamian noise To further engage users, incentive programs and reward tokens will be implemented. Contributors may receive incentives or unique benefits for participating in campaigns. The system will provide enduring methods of engagement and foster a vibrant culture of supporters and creators.

Based on the principles of decentralization, these features will turn the platform into an ecosystem that is scalable, safe, and user-friendly for crowdfunding on the blockchain. Our objective is to address the issues inherent in the conventional approach through the changes outlined below, while preserving the decentralized essence of fundraising..

REFERENCES

1. Junaid Akram and Ali Anaissi, "Decentralized PKI framework for data integrity in spatial crowdsourcing drone services," *Proceedings of the IEEE International Conference on Web Services (ICWS 2024)*, Shenzhen, China, vol. 0087, pp. 643–653, 2024.
2. Shyam Kumar Sinha and Debarka Mukhopadhyay, "Time efficient hash key generation for blockchain enabled framework," *IEEE Access*, vol. 12, pp. 155867–155884, 2024.
3. Hoai-Nam Nguyen, Hoang-Anh Pham, Nguyen Huynh-Tuong, and Duc-Hiep Nguyen, "Leveraging blockchain to enhance digital transformation in small and medium enterprises: challenges and a proposed framework," *IEEE Access*, vol. 12, pp. 74961–74978, 2024.
4. Ziad Hussein, May A. Salama, and Sahar A. El-Rahman, "Evolution of blockchain consensus algorithms: a review on the latest milestones of blockchain consensus algorithms," *Cybersecurity*, vol. 6, pp.1–22, 2023.
5. Pooja Kherwa and Jyoti Arora, "Decentralized social media app on ethereum blockchain," *International Journal of Engineering Applied Sciences and Technology*, vol. 8, pp. 79–83, 2023.
6. Kwame Omono Asamoah, Adjei Peter Darko, Collins Opoku Antwi, Seth Larweh Kodjiku, Esther Stacy E. B. Aggrey, Qing Wang, and Jia Zhu, "A blockchain-based crowdsourcing loan platform for funding higher education in developing countries," *IEEE Access*, vol. 11, pp. 24162–24174, 2023.
7. Weize Xu, Hongyue Duan, Xiao Chen, Jie Huang, Deyong Liu, and Yichao Chen, "Blockchain-based multi-skill mobile crowdsourcing services," *Journal of Wireless Communications and Networking*, vol. 55, pp. 1–27, 2022.
8. Khaled Chait, Abdelkader Laouid, Mostefa Kara, Mohammad Hammoudeh, Omar Aldabbas, and Abdullah T. Al-Essa, "An enhanced threshold RSA-based aggregate signature scheme to reduce blockchain size," *IEEE Access*, vol. 11, pp. 110490–110501, 2023.

9. Kumud Saxena, Vaibhav Kushwaha, Umang Gupta, Vanshika Saxena, and Shweta Srivastav, "Ethereum transaction using meta mask wallet," *International Research Journal of Modernization in Engineering Technology and Science*, vol. 05, pp.7309–7316, 2023.
10. N. Garg, S. Seth, N. Rastogi, R. Kumar, V., Gupta and S. S. Rawat, "Crowdfunding using blockchain technology: a review," *Global Journal of Innovation and Emerging Technology*, vol. 1, no. 2, p. 107, 2022.
11. M. Stefanović, Đ. Pržulj, S. Ristić, D. Stefanović, & and. Nikolić, "Smart contract application for managing land administration system transactions," *IEEE Access*, vol. 10, p. 120897, 2022.
12. I. Malakhov, A. Marin, S. Rossi, and D. Smuseva, "On the use of proof-of-work in permissioned blockchains: security and fairness," *IEEE Access*, vol. 10, p. 120898, 2021.
13. M. Muneeb, Z. Raza, I.U. Haq, and O. Shafiq, "SmartCon: a blockchain-based framework for smart contracts and transaction management," *IEEE Access*, vol. 9, p. 120898, 2021.
14. D. Hawashin, D.A.J. Mahboobeh, K. Salah, R. Jayaraman, I. Yaqoob, M. Debe, and S. Ellahham, "Blockchain-based management of blood donation," *IEEE Access*, vol. 9, p. 120898, 2021.
15. Loan T.Q. Nguyen, T.G. Hoang, L.H. Do, X.T., Ngo, P.H.T. Nguyen, G.D.L. Nguyen, and G.N.T. Nguyen, "The role of blockchain technology-based social crowdfunding in advancing social value creation," *Technological Forecasting and Social Change*, vol. 166, p. 120898, 2021.
16. R.S. Sidiq, A. Jalil, and W.R.W. Achmad, "Virtual world solidarity: How social solidarity is built on the crowdfunding platform Kitabisa.com," *Webology*, vol. 18, no. 1, p.192, 2021.
17. C.K. Upadhyay, B. Pandiya, and V. Tewari, "Digital disbursements over crowdfunding platform for start-ups in India with blockchain technology: A conceptual framework," *International Journal of Blockchains and Cryptocurrencies*, vol. 2, no. 3, p. 263, 2021.
18. U. Khan, Z. Yong An, and A. Imran, "A blockchain Ethereum technology-enabled digital content: development of trading and sharing economy data," *IEEE Access*, vol. 8, pp. 204331–204344, Nov. 30, 2020.
19. H.L. Pham, T. H. Tran, T. D. Phan, V.T.D. Le, D.K. Lam, and Y. Nakashima, "Double SHA-256 hardware architecture with compact message expander for bitcoin mining," *IEEE Access*, vol. 8, pp. 115447–115460, 2020.
20. H. Al-Shaibani, N. Lasla, and M. Abdallah, "Consortium blockchain-based decentralized stock exchange platform," *IEEE Access*, vol. 8, pp. 102455–102468, 2020.
21. T.P. Adewumi, and M. Lewicki, "Inner for-loop for speeding up blockchain mining," *Open Computer Science*, vol. 10, no. 1, pp. 14–27, 2020.
22. National Institute of Standards and Technology (NIST), Secure Hash Standard (SHS), FIPS PUB 180-4, Aug. 2015.

7 Cyberbullying Text Classification Using NLP
Algorithm-Based Techniques

R. Shreya, S. Kolangiammal,
M. Maneesh, and Ankit Gupta

7.1 INTRODUCTION

Numerous adolescents dedicate a significant portion of their time to social networking platforms, where information sharing predominates. Social networks facilitate seamless communication and information dissemination among individuals, regardless of geographical barriers or time constraints. Globally, the user base of social media surpasses 3 billion individuals. According to the National Crime Prevention Council (NCPC), cyberbullying, which involves the deliberate dissemination of hurtful or embarrassing content via mobile devices, gaming applications, or other digital means, poses a prevalent online threat. This form of harassment transcends temporal constraints, occurring incessantly throughout the day and week, and can target individuals anywhere with internet access. Perpetrators of cyberbullying frequently disseminate textual content, photos, or films anonymously, complicating efforts to trace the origin of such content material and impeding its removal. Prominent social media structures, along with Twitter, Instagram, Facebook, YouTube, Snapchat, Skype, and Wikipedia, serve as not unusual venues for cyberbullying incidents. Extensively, platforms such as Facebook provide assets to deal with and prevent cyberbullying, such as devoted sections guiding customers on reporting and blocking offenders. On Instagram, users possess mechanisms to manipulate discomforting content material by way of tracking or blocking unique people while additionally reporting violations to the platform's community recommendations. Social interactions increasingly transcend physical limitations and facilitate unregulated interactions with unknown people, as the net renders them continuously vulnerable to assault with only a click on. The ramifications of cyberbullying amplify past the virtual realm, causing mental, physical, and emotional misery upon victims. Predominantly manifesting as textual or visible harassment on social media structures, cyberbullying may be mitigated through effective detection systems. A talented cyberbullying detection system would determine bullying content from benign textual content, thereby permitting well-timed intervention. Such systems examine numerous aspects of text and appoint contextual cues to determine the underlying meaning. Growing personalized

DOI: 10.1201/9781003567585-7

detection systems capable of effectively and appropriately identifying cyberbullying content remains vital in curtailing the prevalence of online harassment.

7.2 RELATED WORKS

Numerous research has been conducted to deal with the problem of cyberbullying detection, making use of numerous methodologies and ML approaches. M. Di Capua [1] proposed an unsupervised approach method that combines textual and social functions to develop a cyberbullying version. J. Yadav et al. [2] applied a GHSOM network and k-means clustering algorithm to categorize data from the Formspring and YouTube datasets. While their hybrid unsupervised technique yielded promising effects on the Formspring dataset, its overall performance diminished on the YouTube dataset because of variations in textual analysis and syntactical features between the two structures.

Another study via R. R. Dalvi et al. [3] targeted detecting and stopping cyberbullying on Twitter through the use of supervised class ML classifiers, which include aid vector devices and Naive Bayes. With the aid of collecting tweets through the Twitter API and using TFIDF vectorization to extract capabilities, they finished exceptionally high accuracy rates, with the guide Vector gadget outperforming Naive Bayes.

Trana R.E. et al. [4] aimed to limit cyberbullying incidents by designing a machine learning version able to study textual content extracted from photograph memes on YouTube. They compared the performance of three ML algorithms—NB, SVM, and CNN—on the YouTube dataset, with Naive Bayes yielding superior outcomes in unique classes, including race, ethnicity, politics, and generalize.

N. Tsapatsoulis [5] supplied a comprehensive evaluation of cyberbullying on (X) and outlined sensible steps for growing an effective cyberbullying detection application with the use of machine-gained knowledge. The paper discussed trends in data categorization, ML fashions, characteristic kinds, and case studies, serving as a foundational resource for cyberbullying detection projects.

G. A. León-Paredes et al. [6] developed a cyberbullying detection version in Spanish using NLP and ML classifiers like NB, SVM, and LR. Their version executed excessive accuracy quotes and could be further improved by implementing stemming and lemmatization strategies in NLP.

P. K. Roy [7] proposed a DCNN for the detection of hate speech on (X), leveraging ML classifiers together with LR, RF, NB, SVM, DT, Gradient Boosting, and KNN. Their model produced encouraging results, particularly the DCNN variant with 10-fold cross-validation, which demonstrated notably high accuracy in hate speech detection. Ultimately, S. M. Kargutkar et al. [8] developed a cyberbullying detection technique using CNN and Karas, accomplishing an accuracy fee of 87% on Twitter and YouTube records. Their research highlighted the efficacy of deep mastering-based fashions in figuring out cyberbullying episodes with more precision compared to traditional methodologies. The studies by Jamil H. et al. and Rasel et al. [9, 10] reveal the diverse procedures and methodologies used in cyberbullying detection studies, highlighting the importance of leveraging superior machine learning techniques and multidimensional characteristic sets for powerful mitigation of cyberbullying incidents.

Traditional text preprocessing techniques, including stop word elimination, stemming, and lemmatization, exhibit limitations in their capacity to adequately analyze cyberbullying text owing to its informal and evolving characteristics. Current methodologies are inadequate in addressing the challenges posed by slang, abbreviations, typographical errors, and spelling variations, which frequently occur in digital communications (e.g., "u" for "you," "h8" for "hate"). Furthermore, the significance of emojis and emoticons in expressing sentiment cannot be overstated; however, these elements are often overlooked or discarded, resulting in a detrimental loss of contextual significance. This discrepancy underscores the imperative for the development of context-aware preprocessing methodologies to improve the precision of cyberbullying detection.

7.3 PROPOSED METHODOLOGY

Figure 7.1 shows a technique for detecting cyberbullying on social media that goes beyond conventional sentiment evaluation by examining syntactic, semantic, and sarcastic factors of the text. Our method uses contextual mining to gather subjective information from the text in order to determine the reader's viewpoint, sentiment, or attitude toward the issue. Additionally, we introduce a set of "social" features that play a crucial role in guiding the cyberbullying detection process.

Each of these qualities individually identifies the text, and they are categorized according to methods that are already in place in the literature. The efficiency of classifiers in pattern recognition and classification tasks depends on the selection of independent, descriptive, and informative features. When evaluating emotional characteristics. Semantic features focus on identifying lexical relations between words, including trigrams and bigrams, to understand the meaning of the text. We also consider negation and pronoun mapping to account for implicit or explicit references to individuals in harassing messages. Social features delve into the behavioral patterns of both the victim and the bully. Direct tagging of the victim in hate speech, analyzing past confrontations between the bully and bullied, and past malicious activities on social media platforms are considered to provide context to the post and aid in detection. The analysis of multilingual cyberbullying text presents a myriad of challenges, including, but not limited to, language detection, translation, and code-switching phenomena. Models designed for language identification, such as FastText and LangID, possess the capability to accurately discern various languages, whereas machine translation instruments, exemplified by the Google Translate API, facilitate the standardization of text for subsequent analysis. In addressing the complexities of code-switching, the utilization of multilingual embeddings, such as mBERT and XLM-R, coupled with subword tokenization techniques, including Byte Pair Encoding, significantly enhances the processing efficacy of inputs characterized by mixed languages. The incorporation of context-aware natural language processing models serves to further augment the interpretative accuracy of the analysis. The synergistic application of these methodologies contributes to the preservation of both

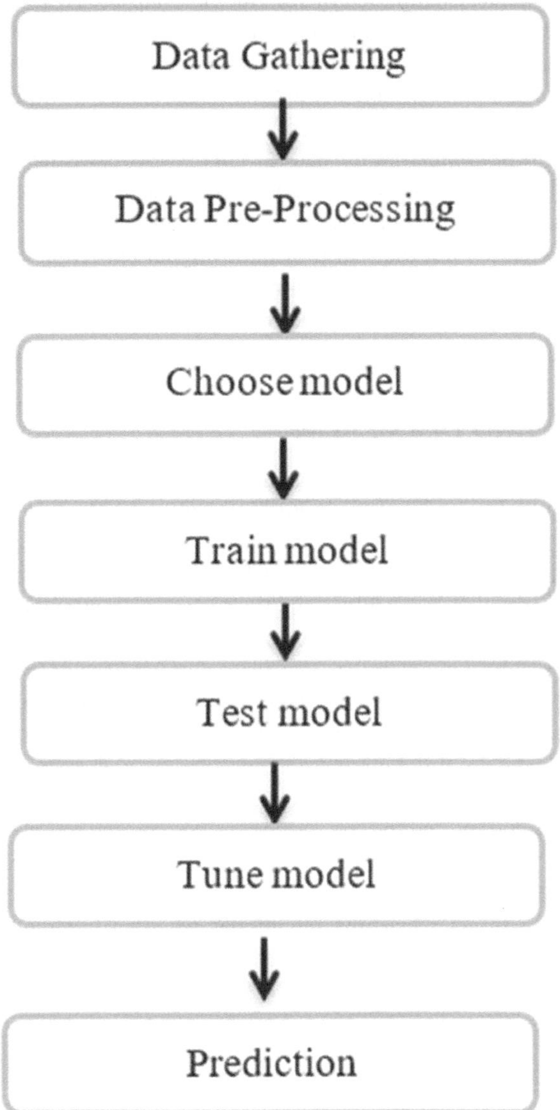

FIGURE 7.1 Flowchart for detecting Cyberbullying on social media.

linguistic and contextual subtleties within the realm of multilingual cyberbullying detection.

The analysis of multilingual cyberbullying text presents a myriad of challenges, including but not limited to language detection, translation, and code-switching phenomena. Models designed for language identification, such as FastText and LangID, possess the capability to accurately discern various languages, whereas machine translation instruments, exemplified by the Google Translate API, facilitate the standardization of text for subsequent analysis. In addressing the complexities of code-switching, the utilization of multilingual embeddings, such as mBERT and XLM-R, coupled with subword tokenization techniques, including Byte Pair Encoding, significantly enhances the processing efficacy of inputs characterized by mixed languages. The incorporation of context-aware natural language processing models serves to further augment the interpretative accuracy of the analysis. The synergistic application of these methodologies contributes to the preservation of both linguistic and contextual subtleties within the realm of multilingual cyberbullying detection. By integrating these features, our method aims to enhance the accuracy and effectiveness of cyberbullying detection on social media platforms, thereby contributing to the development of constructive applications for mitigating cyberbullying issues.

7.4 RESULTS AND DISCUSSION

The text-classified report as a result of testing data, as shown in Figure 7.2, illustrates the performance of our model in distinguishing between bullying (label 0) and non-bullying (label 1) tweets.

In contrast, Figure 7.2 showcases the BERT model's best accuracy on the same Twitter (X) datasets. Remarkably, our BERT-based sentiment analysis model achieved an accuracy of 91.90%, outperforming traditional ML models on similar datasets. These results demonstrate the superior performance of the BERT model in sentiment analysis for cyberbullying detection on Twitter data, with drastically better accuracy attained by using our proposed version underscores its effectiveness in figuring out and categorizing bullying content, thereby contributing to the development of cyberbullying detection methodologies.

- Exploring ensemble methods combining BERT with other models to enhance detection capabilities
- Investigating transfer learning techniques to adapt our model to diverse online platforms and languages

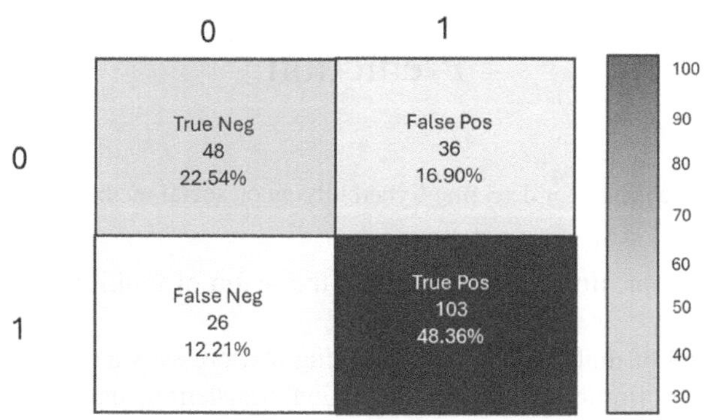

FIGURE 7.2 Visualizing the confusion matrix.

- Developing a comprehensive feature set that captures the nuances of cyberbullying behaviors
- Designing user-friendly reporting mechanisms and intervention strategies to support victims of cyberbullying
- Continuously updating our model to address evolving cyberbullying tactics and online environments

7.5 CONCLUSION

In our research, we have proposed a partially supervised approach method to cyberbullying detection utilizing the BERT version and five key features that represent cyberbullying posts or messages. Especially focusing on sentimental functions, the BERT model achieved an excellent state-of-the-art accuracy of 91%. After training for ninety epochs across two cycles, it outperformed traditional machine learning models. But it's crucial to be aware that the BERT version's accuracy can doubtlessly be enhanced in addition to getting access to a larger dataset. Expanding our method to contain all proposed functions outlined in our studies chapter holds promise for achieving even better consequences in cyberbullying detection. By way of leveraging a comprehensive set of ultra-modern capabilities, we can broaden an application to identify lines of today's cyberbullying and facilitate reporting mechanisms for such posts. Moreover, combining different fashions with the BERT model in future research endeavors can lead to the improvement of the present-day version tailored especially for NLP tasks in cyberbullying detection. Overall, our findings advocate that using advanced strategies together with the BERT model alongside comprehensive function sets can appreciably enhance the efficacy of today's cyberbullying detection structures, thereby contributing to the advent of modern, safer, and more supportive online environments.

REFERENCES

1. M. Di Capua, E. Di Nardo, and A. Petrosino. Unsupervised cyberbullying detection in social networks. *ICPR*, pp. 432–437, doi: 10.1109/ICPR.2016.7899672. (2016)
2. J. Yadav, D. Kumar, and D. Chauhan. Cyberbullying detection using pre-trained BERT model. *ICESC*, pp. 1096–1100, doi: 10.1109/ICESC48915.2020.9155700. (2020)
3. R. R. Dalvi, S. Baliram Chavan, and A. Halbe. Detecting a Twitter cyberbullying using machine learning. *ICICCS*, pp. 297–301, doi: 10.1109/ICICCS48265.2020.9120893. (2020)
4. R.E. Trana, C.E. Gomez, and R.F. Adler. Fighting cyberbullying: an analysis of algorithms used to detect harassing text found on YouTube. In: T. Ahram (eds), *Advances in Artificial Intelligence, Software and Systems Engineering. AHFE 2020. Advances in Intelligent Systems and Computing*, vol 1213. Springer, Cham. (2021).
5. N. Tsapatsoulis and V. Anastasopoulou. Cyberbullies in Twitter: a focused review. *SMAP*, pp. 1–6, doi: 10.1109/SMAP.2019.8864918. (2019)
6. G. A. León-Paredes et al. Presumptive detection of cyberbullying on twitter through natural language processing and machine learning in the Spanish language, *CHILECON*, pp. 1–7, doi: 10.1109/CHILECON47746.2019.8987684. (2019)

7. P. K. Roy, A. K. Tripathy, T. K. Das, and X. -Z. Gao. A framework for hate speech detection using deep convolutional neural network. *IEEE Access*, vol. 8, pp. 204951–204962, doi: 10.1109/ACCESS.2020.3037073. (2020)
8. S. M. Kargutkar and V. Chitre. A study of cyberbullying detection using machine learning techniques. *ICCMC*, pp. 734–739, doi:10.1109/ICCMC48092.2020.ICCMC-000137. (2020)
9. H. Jamil, and R. Breckenridge. Greenship: a social networking system for combating cyber-bullying and defending personal reputation. *ACM* no page. (2018)
10. Risul Islam Rasel, Nasrin Sultana, Sharna Akhter, and Phayung Meesad. Detection of cyber-aggressive comments on social media networks: a machine learning and text mining approach. 37–41, doi: 10.1145/3278293.3278303. (2018)

8 Enhancing Election Integrity with Cybersecurity Solutions

Subhajit Roy, Rajesh Kumhar, Tarkesh Kumar Mahato, Ritwik Priya, Babli Kumari, and Rishabh Dev Shukla

8.1 INTRODUCTION

Electronic voting systems (EVSs) represent a landmark change in the modernization of elections, providing benefits including efficiency, accessibility, and speed [1]. EVSs have changed the landscape of elections by utilizing technology to replace paper ballots with more convenient electronic ballots for voter participation [2]. This also raises concerning issues related to cybersecurity threats which could delegitimize elections overall [3]. Electronic voting refers to modern technologies that include direct recording electronic (DRE) machines, optical scan systems, and online voting, all of these systems could create malicious cyber vulnerabilities [4]. Electronic voting could be more attractive for cybercriminals to compromise [5]. Threat actors can include nation-states, hacktivists, and cybercriminals that can exploit these electronic systems with tactics such as malware, ransomware, phishing, denial-of-service (DoS) attacks, and insider attacks [6].

A major public case was the "hacking" of voter registration systems and voter registration databases found in the 2016 US presidential election that crystallized the impact on public elections today. Similar challenges have arisen in countries such as India, Estonia, and Brazil, and as a result, this is a global issue also [7, 8]. Ensuring proper safeguards for EVSs requires a much larger cybersecurity effort. Strategies for securing elections and EVS include strong encryption, role-based access controls, intrusion detection systems, and having well-articulated incident response plans [9]. New technologies, such as blockchain, have the potential to provide traceability and immutable records. Collaboration within the election ecosystem between governments, cybersecurity experts, and electoral authorities will strengthen opportunities for a coordinated response. Human factors are equally important. While most breaches can be attributed to human error, a surprising number are the result of commonsense mistakes, such as forgetting to follow normal procedures or falling for phishing. Public awareness and education of electoral staff comprise crucial aspects

DOI: 10.1201/9781003567585-8

of cyber defense. International cooperation is also necessary because the threat of cyberattacks reaches beyond borders, in a way that necessitates globally accepted security standards [10, 11].

8.2 CYBERSECURITY LANDSCAPE AND THREATS

Modern technological developments have greatly improved the voting process for voters by making it easier, faster, and more accessible than ever before. A more streamlined process and quicker results are available to voters. We are becoming increasingly aware of serious cybersecurity risks that can limit our ability to protect election integrity, which is concerning since digitally upgrading our electoral procedures and the voter experience holds great promise. Subtle hacking of an EVS has the potential to cause altered results, a decline in public confidence, and problems with the election process as a whole.

8.2.1 Cybersecurity Landscape

The change from traditional paper ballot options to electronic voting, via DRE machines, optical scanners, and Internet voting, has resulted in new possible opportunities for cyber threats. Opponents have utilized increasingly sophisticated methods to exploit even the smallest vulnerabilities present in voting systems. The sources of the threats come from a wide variety of adversaries, including nation-state actors, hacktivist groups, organized cybermortals using ransomware, and all sorts of methods to undermine trust in EVSs. The US national elections in 2016 demonstrated the opportunities for foreign interference and huge risks associated with digital voting systems. The second major risk refers to lack of standardization: voting systems, security protocols, regulatory practices, and more vary from state to state and from one local jurisdiction to another, creating vulnerability gaps and making cyberspace response to a cyber incident much more difficult [12].

8.2.2 Types of Cyber Threat Vectors

There is a wide range of threats to election processes and electronic voting systems that impact a democratic process in cyberspace with varying degrees of severity. Malware attacks include malicious code that locks up election data to continue exfiltrating until the ransom is paid or trojan horses embedding malicious code to alter vote counts or extract sensitive information [13].

8.2.3 Denial-of-Service (DoS) Attacks

This occupies another layer in that they use the available computer market to saturate the election system with horrendous amounts of traffic so they cannot function properly during a critical voting time period. While DoS attacks would not necessarily damage the data, they impact even more operational processed and voter confidence [13].

8.2.4 Insider Threats

It is possible for trusted people with access to systems to be intentionally or inadvertently made vulnerable. Carelessness from employees not covering the basics of security, such as not following security protocols, also introduces significant risks.

Phishing and Social Engineering Attacks: Let us expand on the types of attacks threatened to election processes at the level of human error. Hackers can act as trusted sources to induce election workers to share sensitive information or to click a malicious link. Spear phishing would be the most dangerous for elections as it is a more easier way of targeting a person.

Network Vulnerabilities: Weak and unprotected networks could provide access through man-in-the-middle attacks or unauthorized access from bad actors trying to gain entrance to an unprotected system, especially if that system is connected via public Wi-Fi or once again lacks security.

8.2.5 Hack-and-Leak Operations

Cybercriminals could steal and leak data from a campaign and then use that information to manipulate public awe and damage a candidate's credibility [14].

8.2.6 Supply Chain Attacks

Attackers could use third-party vendors or create unintentional backdoors in the firmware or hardware components of user systems, as happened in SolarWinds.

8.2.7 Disinformation and Misuse of AI

Deep Fakes and misleading AI generative content can spread virally, leading to manipulating trust and perception on the part of voters.

We must recognize and address threats [15] in advance, looking to protect the election process and civil democracy in an increasingly digital environment shown in Table 8.1.

8.2.8 Consequences of Cyberattacks on EVSs

Cyberattacks on EVSs can have considerable impacts beyond mere technical failures. The most potentially damaging effect of a cyberattack may be the erosion of public trust. Even the perception of a compromise to a voting system, even without an actual attack, can lead to minimal engagement at the polls and diminished confidence in the democratic process. If an attack to manipulate votes does occur, the legitimacy of elected governments may be undermined, along with the legitimacy of democratic governance in general. Beyond political ramifications, breaches of security typically involve legal disputes, regulatory investigations, and reputational damage. For example, election officials may be sued or publicly criticized for not

TABLE 8.1
Digital Election Threat Pathways

	Threat Vector	Election Campaigns and Voters	Election Administrators	Election Systems
Cyber intrusion	Unapproved access and information theft	●	●	○
	Malicious actions by insiders	○	○	○
	Manipulation of voting systems	—	○	○
Hybrid	Disruptive cyberattacks	○	○	○
	Distributed Denial of Service	●	●	○
	Website alterations	○	○	—
	Hack and leak operations	●	●	○
	Cyber actions inciting real-world protests	●	○	○
	Direct outreach to targets via email, SMS	●	○	—
IO	Boosting of influence operation content on social media	●	—	—
	Sharing of influence operation materials through websites	●	—	—
	Exploitation of AI technologies	●	—	—

protecting electoral infrastructure, raising important concerns about accountability and transparency.

These incidents serve as a clarion call in the importance of cybersecurity to mitigate cyberattacks on EVSs. Such scheme helps protect democracy in an increasingly digital society.

8.3 CYBER THREATS TO ELECTION INTEGRITY AND GLOBAL CASE STUDIES

Election systems globally are navigating a fast-growing landscape of cyber threats. These threats reveal that there are critical vulnerabilities in our EVSs and, in many cases, expose a lack of cybersecurity strategies that provide a coordinated and robust response to cyberattacks. Our adversaries continue to develop increasingly complex ways to disrupt, manipulate, or otherwise de-legitimate any election process. The following examples illustrate the global scale and increasingly sophisticated nature of these attacks.

US 2016 Presidential Election: Russian operatives, allegedly, hacked Jacksonville Florida's state election systems to gain access to state voter registration databases and other critical election infrastructure in the November 2016 presidential election.

While they did not change any votes, it raised some serious concerns about the integrity of the elections and the need to develop better cybersecurity. Russian state actors, including APT28 and Sandworm Team (APT44), continue the hybrid operations targeting elections today In 2019, a security analyst revealed serious vulnerabilities in Georgia's outdated EVS, demonstrating that a remote hack could, in fact, change vote counts. This was a critical expose of the need for updating systems regular security testing, and access to the network. Networks with known vulnerabilities encourage cybercriminals, nation-state actors, and possibly rogue/terrorist entities.

Venezuela—Mid April 2018: The attentions of outside hackers disrupted the elections in Venezuela in 2018 when hackers took aim at voting systems to delegitimize the election. Although they did not alter any election outcomes, the hackers—who were tied to Russian IR6/IR7 and Iranian IR5—illustrated vulnerabilities amidst a political chaos. The hackers combined cyberattacks and information operations to erode public trust.

Estonia: Online Voting Under Attack (2007): Estonia was an early adopter of online voting. Massive cyberattacks against Estonia, which were promoted by Russian political agendas, forced Estonia to completely re-vamp its cybersecurity architecture. The attack helped lead to establishing NATO's Cooperative Cyber Defence Centre of Excellence. Estonia's actions reflected the vulnerabilities of online voting and the need to have secure tectonic frameworks in stable societies and regions.

Brazil: Cybersecurity Vulnerabilities for the Election (2018): Brazil experienced breaches involving election-related data during its 2018 elections. The breaches did not affect vote integrity but since the experience showed the importance of continuous monitoring and updating. Brazil then rolled out multifactor authentication (MFA), strengthened its network segmentation, and implemented real-time monitoring.

Coffee County, Georgia: Insider Risk (Unauthorized Access) (2021): Reporting of unauthorized access in Coffee County, Georgia took place following the controversial 2020 US presidential election. Sensitive election software and data were copied, underscoring risks with insider threats and election technology. Similar risks come from groups such as Iranian APT42 and UNC2248, which target infrastructure for cyber espionage and influence campaigns.

Cyberattack on a US county (2023): In 2023, a ransomware incident impacted a US county election system that was down for an extended period of time. The officials had to operate under extraordinary circumstances to secure a new system. This incident underscored what ransomware could do to elections, especially relative to the timing with elections. In addition to ransomware, however, cybercriminals are increasingly using extortion tactics that may influence elections. Others (either non-state or state actors enabling pro-China influence) have engaged in election actions around the world, generating disinformation related to elections intended to shape voter's perceptions and confound voters, particularly in Taiwan.

Russian Threat Groups: Russian state-sponsored cyber actors such as APT28, APT44 (Sandworm Team), UNC4057 conducted operations that impacted elections all over the world. Their actors merge cyber espionage, disruptive technical effects, and information operation (IO) activities in a manner directed to engender mistrust in elections, as discussed with the 2016 hack-and-leak of the US election and the 2016 EU referendum. Russian threat actors have been implicated in attacks launched against Ukraine, France, and Georgia by combining DDoS activity, data leaks, and disinformation to create distrust.

Chinese Influence Operations: Chinese threat actors have stepped up their election influence efforts around the world via information operations. DRAGONBRIDGE and HaiEnergy, for example, integrate AI-synthesized deepfakes and impersonate legitimate organizations to amplify pro-China narratives and attack opponents. In December 2023, APT42 used fake news and misleading content to influence public opinion during Taiwan's 2024 presidential election.

Iranian Influence Operations and Espionage: Iranian cyber groups (primarily APT42 and UNC2248) typically target elections with cyber espionage, online narrative pushes, and voter intimidation or influence. UNC2248 has targeted and exploited vulnerabilities to compromise US government organizations, while APT42 has used credential harvesting to secure sensitive data for their upper hand in elections. Their narratives are primarily anti-US and anti-Israel, and they have had lots of activity whenever there is a US election cycle.

North Korean cyber actors are increasingly targeting election-related organizations with espionage, rather than directly manipulating votes. APT43 and APT31 followed through on phishing and spearphishing campaigns targeting government organizations, political actors, and election stakeholders ahead of the 2020 US election. Their intent was to harvest sensitive information—emails, strategic plans, and personnel files—to better strategize on their political goals. While they had no intention of changing cast ballots, these operations showcase the natural combination of hybrid attacks grouping cyber espionage and information warfare, exploiting human weaknesses to create distrust by undermining confidence and the integrity of electoral processes. This type of activity demonstrates the evolving threat environment for electronic voting systems (EVSs). Indirect intrusions can be just as dangerous as overt attacks.

As these examples show, electoral systems must confront increasingly sophisticated threats. A holistic defense—standardizing security protocols, vetting supply chains, building resilient architectures, training personnel against phishing and insider risks, and sharing threat intelligence internationally—is essential to maintain secure, verifiable, and trusted elections, shown in Table 8.2.

8.4 DEFENSE MECHANISMS FOR SECURE ELECTRONIC VOTING

Since cyber threats can exploit vulnerabilities in many ways, safeguarding EVSs requires implementing an overall, layered defense that incorporates technical, administrative, and physical measures to protect the integrity and confidentiality of the voting process. The following sections outline important security mechanisms for defending against cyber threats to EVSs [16, 17].

TABLE 8.2
Active Threat Actors across the World

	Threat Actor	Likelihood of Activity	Type of Activity	Hybrid
Russia	APT44	High	Intrusion, Hybrid, IO	Capable
	UNC4057	High	Intrusion, Hybrid, IO	Capable
	KillNet and similar hacktivist groups	High	Hybrid	Capable
	NAEBc and other IO campaigns	High	IO	—
	APT28	Moderate	Intrusion, Hybrid, IO	Capable
	UNC2589	Moderate	Intrusion, Hybrid, IO	Capable
	APT29	Moderate	Intrusion	—
Iran	APT42	High	Intrusion	—
	Pro-Iran Hacktivists	Moderate	Hybrid, IO	Capable
	IO Campaigns	High	IO	—
	UNC757	Moderate	Intrusion, Hybrid	Capable
	UNC2448	Low	Intrusion, Hybrid	Capable
China	TEMP.Hex	High	Intrusion	—
	APT41	Moderate	Intrusion	—
	APT31	Moderate	Intrusion	—
	APT40	Moderate	Intrusion	—
	UNC3658	Moderate	Intrusion	—
	HaiEnergy	High	IO	—
	DRAGONBRIDGE	High	IO	—
	Fictitious Brands	High	IO	—
North Korea	APT43	Moderate	Intrusion	—

8.4.1 Data Encryption

Encryption makes sensitive information unreadable to unauthorized parties. Data in-transit—such as moving vote records from machines to central servers—should use a secure transfer protocol such as Transport Layer Security (TLS), making the data unreadable if intercepted. Data-at-rest—vote tallies or registration databases on the machines or servers—should also be encrypted so that if an attacker were to steal drives or gain unauthorized access to stored information, they cannot read or change that data.

8.4.2 Access Control

Role-based Access Control (RBAC) limits functions of the system based on the user's role (e.g., only election officials can change tallies, while pollworkers can only check in voters). MFA adds a layer of protection because additional proof is required for entry (e.g., a one-time code or hardware token); therefore, stolen credentials can't just provide access to critical systems.

8.4.3 Auditing and Monitoring

Auditing and monitoring with continuous Intrusion Detection Systems (IDS) ensures malicious traffic on the network is identified and alerted in real-time, allowing you to promptly react. Log management tools must log all access experiences and configuration changes. If officials review logs regularly to look for anomalies—such as failed logins and unusual data exports. They will have the opportunity to investigate suspicious behavior before it becomes a full breach.

8.4.4 Incident Response Plans

An incident response plan defines roles, responsibilities, and procedures for detecting, containing, and remediating security incidents. Incident response plans should specify communication channels, escalation paths, and indications of recovery for different attack types. Regular drills and tabletop exercises ensure some level of familiarity among personnel so that if a real event occurs, their actions can be swift.

8.4.5 Physical Security Controls

Physical security controls are the buying force of a security program. Physical security controls thwart tampering of machines and data. Strong physical security controls include locked access-controlled rooms with guards and video surveillance, where you can store voting devices and voting media. Tamper-evident seals that are placed on voting machines mean that someone opened the machine and must be investigated while properly documenting it. The choice of appropriate locks and entry logs is strong public policy to help protect voting media and devices from tampering by insiders (and also protects chain-of-custody).

8.4.6 Software Security

Keeping software updated and patched eliminates known vulnerabilities in voting software. Secure coding practices involving input validation, error handling, and least-privilege design can aid in reducing the possibility of new vulnerabilities being introduced. Code reviews, unit testing, and penetration-testing approaches will expose weaknesses prior to being deployed into the production environment so that necessary patches can be put into place to help reject cyberattackers from exploiting those weaknesses.

8.4.7 Periodic Security Audits

Independent third-party audits provide unbiased assessments of an EVS's security posture. Audits test compliance with standards (e.g., NIST guidelines), evaluate the effectiveness of controls, and uncover hidden gaps. Compliance reviews and penetration tests should occur before and after each election cycle, with remediation tracked to ensure continuous improvement.

By implementing this layered suite of defenses—encryption, strict access controls, vigilant monitoring, practiced response planning, physical safeguards, secure software development, stakeholder training, collaborative intelligence-sharing, and regular audits—election officials can significantly reduce the risk of cyber compromise. These measures work together to preserve public confidence, uphold electoral integrity, and ensure that every vote remains private, accurate, and tamper resistant.

8.5 COMPREHENSIVE PROTECTION STRATEGIES FOR ELECTRONIC VOTING

The protection of EVS against cyber threats is essential for maintaining the integrity of elections. A strong defensive posture entails a combination of policies, technologies, and practices aimed at reducing exploitable vulnerabilities and mitigating emerging risks. There are several practices, arranged thematically below, that relevant authorities can employ to protect EVS [18, 19].

8.5.1 Risk Assessment and Management

Regular risk assessments are the foundation of any cybersecurity approach. Risk assessments will help EVS officials identify what threats they should be most concerned about and where there are exploitable vulnerabilities.

Comprehensive Risk Assessments: EVS are subject to technical and operational assessments. Technical assessments include penetration tests, vulnerability scans, and red team assessments of expected patterns of attacks based on real-world circumstances. Operational assessments of personnel and admin controls and physical controls to identify weaknesses in daily procedures.

Risk Mitigation Strategies: When the risks are identified, it is up to officials to determine the appropriate responses to the risks identified. Responses can be as simple as replacing or patching old hardware with more recent versions, improving physical barriers, expanding employee awareness training, or allowing segmentation of networks to limit exposure of key service components. Each option for mitigating risk should be implemented and/or prioritized in accordance to the likelihood and impact associated with the threat in question.

8.5.2 Development of Security Policies

Security policies establish the framework that guides every aspect of EVS governance. They define roles and responsibilities that staff must follow.

Comprehensive Security Framework: A good policy covers a lot of ground. A policy should address everything from data protection, access control, incident response, acceptable use, and more. The policy should refer to best standard practices—that is, with respect to NIST guidelines, the agency can update the policy as threats and technology regulations change over time.

Regular Policy Reviews: Cyber threats evolve quickly, so security policies must also evolve. Routine reviews of the policies—at least once per year, and whenever important technological changes occur—will ensure policies are up to date. Policy reviews should be discussed with all stakeholders. Updates should also include sessions to explain any new or changed procedures.

8.5.3 Training and Awareness Programs

Human error remains one of the leading contributors to security incidents. Therefore, instituting a culture of security awareness is an essential step.

Cybersecurity Training: Election officials and poll workers should receive continuous training on issues from identifying phishing schemes, to password hygiene, to safely and securely handling election data, to correctly using devices. Live exercises, such as conducting phishing exercises or attempting unauthorized access, can reinforce the key lessons.

Public Awareness Campaigns: When citizens understand that there are safeguards taking place to protect their data, their confidence in the electoral process increases. A variety of awareness campaigns, from social media posts to informational brochures at polling sites, can describe the various security measures in layman's terms to assure voters they are taking action to protect their data and offer guidance on how to identify official communications.

8.5.4 Incident Response Planning

No matter how much prevention is built into a risk management plan, there is always the possibility that breaches will happen. Having a clear, tested incident response plan ensures that when adversaries strike, the response team will respond quickly and collaboratively.

Craft an Incident Response Plan: An effective Incident Response Plan should outline detection protocols, containment procedures, communication instructions, escalation expectations, and recovery processes. The plan must also assign roles and responsibilities for who makes the decision about backups and calls law enforcement and who speaks to the public when required.

Regular Drills and Testing: Periodic tabletop and live exercises will familiarize personnel with their responsibilities and help identify weaknesses in the plan. Documenting observations and lessons learned from post-exercise debriefs allows the team to improve how to respond to incidents.

8.5.5 Collaboration and Information Sharing

Cyber threats to elections do not acknowledge organizational or national boundaries. The strength of collaboration is to use collective knowledge in order to enhance defenses.

Information Sharing Networks: Election officials can take advantage of membership in information-sharing networks at regional and global levels, gaining insight into threats and timely alerts regarding new vulnerabilities and attack techniques. Sharing sanitized incident data with peers will help a community dependably strengthen its defenses in a purposeful way.

Partnerships with Cybersecurity Practitioners: Hiring outside resources, including those from government, academia, or the private sector, brings diversity and specialized skills in vulnerability assessments, threat hunting, and tailored security architecture. They can also offer support with red team engagements and forensic investigations.

8.5.6 Innovative Security Solutions

Utilizing the latest security technologies will further the technical protections against EVS attacks by making attacks harder and/or slower to attack, while at the same time making unauthorized intrusion easier to detect and respond to.

- MFA: MFA should be a requirement for accessing any critical systems. By requiring two or more verifications using "something you know" (password), "something you have" (hard token or SMS code), or "something you are" (biometric), we raise the bar considerably for any unauthorized intruder.

Intrusion Detection and Prevention Systems (IDPS): IDPS solutions monitor network and system activities to identify signs of malicious behavior. They can then automatically block the suspicious traffic, alert administrators about suspicious activity, and even isolate compromised segments to prevent lateral movement within the network.

Blockchain Technology: While not a silver bullet, blockchain can provide an immutable audit of vote activity. Changes to advisory votes require consensus and an immutable record, so when access is attempted to alter votes, the new record creates a discrepancy that is easily obtainable from the tamper-resistant ledger visible to auditors who have security credentials.

8.5.7 Regular Security Audits and Compliance Reviews

Regular verification and auditing of security controls are essential to sustain a robust security posture over time.

Third-Party Audits: Engaging an independent and qualified cybersecurity firm can provide a fresh perspective and a level of neutrality. Auditors examine the technical configurations, ensure adherence to policy, confirm operational state, and provide the audited entity with a score that reflects their audit scope and tiered recommendations to resolve non-compliance.

Compliance Audits: Elections must adhere to relevant standards—such as the NIST Cybersecurity Framework or ISO 27001. Regular compliance checks verify that implemented controls meet regulatory requirements and internal mandates, helping avoid costly fines or public backlash.

By integrating these defensive strategies—comprehensive risk management, clear policies, continuous training, rapid incident response, collaborative information sharing, advanced technologies, rigorous audits, and a culture of security—election authorities can significantly reduce the risk posed by cyber threats. Together, these measures protect the confidentiality, integrity, and availability of EVS, reinforce public confidence, and uphold the democratic process.

8.6 CONCLUSION

Trust in democratic systems depends upon the cybersecurity of EVSs, as they are the foundation for elections around the world. EVS face various cybersecurity threats, including malware, denial-of-service (DoS) attacks, insider threats, and phishing—threats that severely undermined trust in the 2016 US presidential election and in voting systems in Georgia. Our objective is to instill public trust through cybersecurity and credible security protocols. Cybersecurity requires appropriate risk assessments, updated security tools, and ongoing cybersecurity awareness with all stakeholders. Ensuring the cybersecurity of polling locations, in particular, is essential because a compromise can alienate the public trust or, in a worst-case scenario, destabilize the political system. Future research needs to be developed on security assurance through the use of emerging technologies such as blockchain, the potential of creating standard cybersecurity policies across platforms, and how to increase faith in the system with public engagement. As the threats evolve, there is no substitute for strong, adaptive cybersecurity capabilities, which are needed to defend EVS and the democratic process.

REFERENCES

1. Adams, A.K., 1, A. Azameti 1, G. Koi-Akrofi 1, S.C. Quist 1, R. Dayie 1, and University of Professional Studies. 2024. Systematic Literature Review: A Novel Framework for Evaluating Electronic Voting Systems Artifacts. *EAI Endorsed Transactions on Scalable Information Systems*. Vol. 1. https://adamsblessed7.github.io/azameti/Manuscript_3.pdf.
2. Ajayi, O.A. 2024. *Internet Technologies and Cybersecurity Law in Nigeria*. African Books Collective.
3. Sengupta, S., and S. Roy. 2024. Artificial Intelligence (AI) in Cybersecurity: A Revolution in Threat Detection and Prevention. *International Ethical Hacking Conference 5th*. Kolkata, India. https://doi.org/10.1007/978-981-97-8457-8_31.
4. Appel, A.W., R.A. DeMillo, and P.B. Stark. 2020. Ballot-Marking Devices Cannot Ensure the Will of the Voters. *Election Law Journal Rules Politics and Policy* 19, no. 3: 432–450. https://doi.org/10.1089/elj.2019.0619.

5. Basak, B. 2024. The Impact of Cybersecurity Threats on National Security: Strategies. *International Journal of Humanities Social Science and Management (IJHSSM)* 4 (Mar.): 1361–1382. https://ijhssm.org/issue_dcp/The%20Impact%20of%20Cybersecurity%20Threats%20on%20National%20Security%20%20Strategies.pdf.
6. Breedon, K. and Bryant, A. C. 2019. Counting the Votes: Electronic Voting Irregularities, Election Integrity, and Public Corruption. (Jun. 15). *University of Memphis Law Review* 49, no. 4. https://papers.ssrn.com/sol3/papers.cfm?abstract_id=3784122
7. Buchyn, M., A. Helesh, and B. Shubyn. 2021. Information Security during Electronic Voting: Threats and Mechanisms for Ensuring. *2021 IEEE 4th International Conference on Advanced Information and Communication Technologies (AICT)*, (Sep. 21): 266–269. https://doi.org/10.1109/aict52120.2021.9628971.
8. "Cybersecurity Framework | NIST," NIST, Dec. 05, 2024. https://www.nist.gov/cyberframework.
9. "Election Security | U.S. Election Assistance Commission." https://www.eac.gov/voters/election-security.
10. Hyiamang, O. 2022. How can Cybersecurity Best Practices Protect Election Integrity in Advanced and Developing Democracies? - ProQuest. https://www.proquest.com/openview/2059a8b870d5a479ec66d2be9773b6f8/1?pq-origsite=gscholar&cbl=18750&diss=y.
11. Jafar, U., M.J.A. Aziz, and Z. Shukur. 2021. Blockchain for Electronic Voting System—Review and Open Research Challenges. *Sensors* 21, no. 17 (August 31): 5874. https://doi.org/10.3390/s21175874.
12. Mandiant, 2024. *Poll Vaulting: Cyber Threats to Global Elections.* Google Cloud Blog, (Apr. 25), https://cloud.google.com/blog/topics/threat-intelligence/cyber-threats-global-elections.
13. Risnanto, S., Y.A. Rahim, O. Mohd, N. Kusmadi, A.E. R, and R.S. Perdana. 2021. E-Voting: Security, Threats and Prevention. *2021 15th International Conference on Telecommunication Systems, Services, and Applications (TSSA).* Bali, Indonesia: IEEE. https://doi.org/10.1109/tssa52866.2021.9768214.
14. Shires, J. 2019. Hack-and-Leak Operations: Intrusion and Influence in the Gulf. *Journal of Cyber Policy* 4, no. 2 (May 4): 235–256. https://doi.org/10.1080/23738871.2019.1636108.
15. Stoddart, K. 2022. *Cyberwarfare: Threats to Critical Infrastructure.* Palgrave Macmillan.
16. Theodorakopoulos, L., A. Theodoropoulou, and Y. Stamatiou. 2024. A State-of-the-Art Review in Big Data Management Engineering: Real-Life Case Studies, Challenges, and Future Research Directions. *Eng—Advances in Engineering* 5, no. 3 (July 3): 1266–1297. https://doi.org/10.3390/eng5030068.
17. Toygür, I., W. Quiviger, C. Malmström, M. Muñiz, M. Amand, O. Belogolova, A. Broughel, et al. 2024. *Beyond Electoral Uncertainty.* https://shs.hal.science/halshs-04664450/.
18. Van Der Staak, S., and P. Wolf. 2019. *Cybersecurity in Elections: Models of Interagency Collaboration.* https://doi.org/10.31752/idea.2019.23.
19. Zelizer, J.E., and K.J. Greenberg. 2024. *Our Nation at Risk: Election Integrity as a National Security Issue.* NYU Press.

9 Mitigating Privacy and Security Risks of Federated Learning in Smart Healthcare

Analysis

Ashok Kumar Pradhan, Swetha Ghanta, Subhra Prosun Paul, and Jordan Orange

9.1 INTRODUCTION

A branch of machine learning called FL, or collaborative learning, focuses on scenarios where a number of organizations (commonly referred to as clients) work together to train a model while maintaining the decentralization of their data. This contrasts with settings for machine learning, where data is kept centrally. Basically, a central model is trained across decentralized devices or servers through the process of FL [1]. The model is trained locally on each device, and only the model changes are communicated, as opposed to sending all data to a single place. By not disclosing raw data, this preserves privacy and permits collaborative learning [2]. Data heterogeneity is one of the key elements that characterizes FL. FL seeks to train machine learning algorithms, such as deep neural networks, in which many local datasets may be present in local nodes without requiring explicit data sample exchanges [3]. The fundamental idea is to train local models on local data samples and then exchange parameters (such as a deep neural network's weights and biases) at regular intervals between these local nodes to create a global model that is shared by all nodes [4]. Federated learning can be classified into three main categories: centralized, decentralized, and heterogeneous federated learning.

9.1.1 Importance of FL

In situations such as intelligent health care networks, FL has a number of benefits. First, by eliminating the need to centrally store sensitive data, it makes it possible to train machine learning models across dispersed devices or servers, protecting

 DOI: 10.1201/9781003567585-9

user privacy and lowering the possibility of data breaches. Second, in order to create more accurate and relevant models for specific people or devices, FL enables personalized model updates based on local input. Thirdly, it may more effectively utilize bandwidth and be implemented in contexts with limited resources by drastically lowering the quantity of data that must be sent to a central server. Furthermore, FL allows ongoing learning and adaptation to local data modifications, guaranteeing that models stay current and functional throughout time. FL is significant for the following reasons:

Privacy: FL permits local training on the edge device, avoiding possible data breaches, in contrast to conventional systems that send data to a central server for training [5].

Data security: The central server only receives the encrypted model changes, which guarantees the safety of the data. It is also possible to decode only the aggregated results using secure aggregation techniques such as secure aggregation principle [6].

Heterogeneous Data Retrieval Issue: Unified learning ensures that heterogeneous data dispersed across many devices, places, and organizations is accessible. While preserving confidentiality and privacy, it enables the training of models on sensitive data, like financial or medical data. More diverse data also allows for the generalization of models [7].

When individual players need to train models on datasets bigger than their own but are unable to share the data with others (for legal, strategic, or financial reasons, for example), FL is usually used. The technique still needs the minimum processing power needed by each node and strong connectivity between local servers.

9.1.2 How FL Works

The central server has a general baseline model saved on it. The client devices receive copies of this model, which they use to train their models using the local data they provide. The models on different devices improve in usability and personalization over time [8]. Using safe aggregation techniques, the primary model at the central server receives the updates (model parameters) from the locally trained models in the subsequent step. To provide new learnings, this model averages and mixes several inputs. There is more potential for the model to become broadly applicable because the data is gathered from a variety of sources [9].

The central model is shared with the client devices once more for the subsequent iteration after it has been retrained using the new parameters [10]. The models continue to develop without compromising privacy with each cycle, collecting a diverse range of data. Federated systems, by eliminating the need for large data storage systems and by matching data across different sources, can deliver more accurate information than other forms of databases, making them superior for data integration and analytics [11].

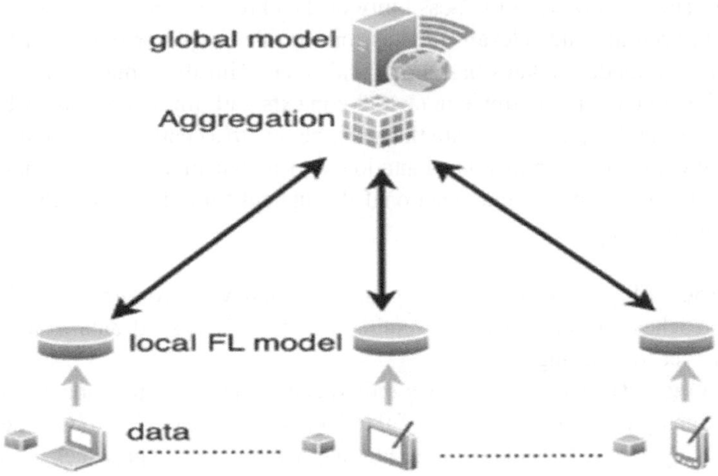

FIGURE 9.1 Logical architecture of FL.

Two main issues with data analysis are resolved by FL: better qualitative assessments for society and protection of people's private rights.

Figure 9.1 shows the logical architecture of the FL working system. In this figure, data is directly supplied to the local FL model; then, through this data aggregation process, this data is transferred to a global model. Large-scale data analysis is much simpler than in the past. Algorithms are becoming more complex and have greater computing power. There is a growing number of societal objections to the use of sensitive data, despite the fact that more and more data is available for insightful studies. Utilizing data without compromising privacy is possible because of FL. Since data may be studied from numerous databases, there is a higher amount of data available. A study's findings are therefore more trustworthy. This implies improved models and forecasts, which translate into far better-informed (policy) choices. Considering the field of cancer research, FL allows for the analysis of anonymized data from all hospitals, preserving patient privacy, on topics such as effective treatment strategies for various cancer types in different individuals [12].

9.1.3 Applications of FL

In situations where training data is dispersed and privacy-sensitive applications need to be supported, FL is crucial. In theory, this seems like a great strategy to address issues with conventional machine learning models that use data from a single central server or location [13]. While there is still much to learn about FL, there are some uses for it currently, such as smartphones, commercial organizations, IoT, healthcare, advertising, autonomous vehicles, and insurance sectors. FL applications have their problems, much like any other machine learning approach. It may outweigh the disadvantages and revolutionize a number of sectors [14]. FL and all of its uses would advance significantly in the near future. Businesses would be happy with a dispersed learning approach. It will enable prompt and cost-effective reactions to rapidly evolving customer behavior.

FL is a very popular method. When used practically, its use may aid in the evolution of several sectors and also provide benefits to the users [15]. Here are some significant applications of FL in commercial sectors:

9.1.3.1 Smartphone

Apps such as voice recognition, facial recognition, and next-word prediction are powered by statistical models that learn user behavior over a large pool of mobile phones. For privacy reasons, or to save on bandwidth or battery life, users can choose not to share their data. Without disclosing private information or degrading user experience, FL may produce precise smartphone predictions.

9.1.3.2 Organizations

In the context of FL, entire institutions or organizations may be seen as "devices". Predictive healthcare apps can benefit from the vast quantity of patient data that hospitals, for instance, hold. However, hospitals have strict privacy regulations in place and may be constrained by administrative, legal, or ethical constraints that mandate local data storage. Because it lowers network burden and allows private learning across several devices and organizations, FL is an excellent fit for these applications.

9.1.3.3 Internet of Things (IoT)

Real-time data collection and response are facilitated by sensors in contemporary IoT networks, which include wearable devices, self-driving automobiles, and smart homes. An updated traffic, construction, or pedestrian behavior model, for instance, could be necessary for a fleet of autonomous vehicles to function properly. Because of privacy concerns and each device's limited connectivity, it could be challenging to create aggregate models in these situations. Train models that adapt to these systems' changes effectively while protecting user privacy thanks to FL techniques.

9.1.3.4 Advertising

We are aware that the data pertaining to each individual user is crucial to the process of personalization. But websites, social media networks, e-commerce platforms, and other places spring to mind when more people worry about how much information they would rather not share with external parties. With its reliance on customer personal data, advertising may employ FL to stay afloat and allay people's fears.

9.1.3.5 Autonomous Vehicles

The development of self-driving cars uses FL because it can produce predictions in real time. Data can include real-time updates on traffic and road conditions, facilitating quicker decision-making and ongoing learning. As a result, using a self-driving car might be safer and more pleasurable. Implementing federated machine learning has great potential in the automotive industry. But all that's being done in this area right now is study. FL has been shown in one study to reduce training time for wheel steering angle prediction in self-driving cars.

9.1.3.6 FL in Financial Fraud

The digital age has led to the emergence of several international financial crimes. Common subcategories of financial crimes include money laundering, fraudulent lending, and financial crimes. Credit card fraud costs banks and customers a lot of money.

9.1.3.7 FL in Insurance

Creating a data service platform for the insurance industry necessitates integrating financial, medical, and other data from several sources. If an insurance company wants to improve its ability to manage risk and develop its business, it needs to take multi-party data into account. In the insurance industry, efficient data use without violating individual privacy is also a major challenge.

9.2 RELATED WORKS

Due to its benefits in protecting the privacy of isolated data, FL has attracted more attention in the past several years for a variety of smart medical applications. This related work section discusses privacy and security considerations in FL for smart healthcare systems.

In-depth discussions of the encryption protocols, safe aggregation, and differential privacy methods built into FL are provided by Bala et al., who highlight the importance of these processes in protecting data privacy. Through an analysis of real-world deployments and case studies, they highlight the usefulness of FL in the field of healthcare informatics. The study also discusses potential difficulties, ethical ramifications, and legal issues related to the application of FL in healthcare settings [16].

Recent developments in FL for smart healthcare are discussed together with the needs and incentives for its implementation in a thorough review by Dinh et al. The survey covers several FL ideas, such as resource-aware FL, incentive FL, secure and privacy-aware FL, and personalized FL that are suited for smart healthcare. It also examines new uses for FL in important areas of healthcare, including medical imaging, remote health monitoring, health data management, and COVID-19 detection. The study outlines research problems and future prospects for FL research in healthcare, analyzes previous FL-based smart healthcare projects, and highlights important lessons learned [17].

A framework for protecting IoT healthcare data privacy through FL and blockchain technology is put out by Saurabh et al. Using blockchain-based IoT cloud platforms for security and privacy, their work presents a Blockchain and FL-enabled Secure Architecture for Privacy-Preserving in Smart Healthcare. Users can use well-trained machine learning models without sending sensitive data to the cloud thanks to the framework's adoption of FL technology for scalable machine learning applications in the healthcare industry. The study also covers FL applications in distributed secure environments in smart cities, highlighting the importance of cryptographic methods like hashing, encryption, and signatures in safeguarding sensitive medical data. A comparative analysis shows how effective their suggested model is in maintaining privacy, being scalable, and having fewer overheads [18].

Haya et al. proposed a Deep FL framework for decentralized healthcare systems to maintain user privacy in a distributed architecture. The paper introduces an algorithm for an automated training data acquisition process and presents experiments demonstrating the application of deep FL in detecting skin diseases. Transfer learning is employed to address the limited availability of healthcare data for building deep learning models. The results indicate a significant improvement in model performance, with FL increasing the Area under the Curve of the centralized learning model up to 0.97. Although the FL system affects the quality of service in terms of model conversion time, it meets the requirements of building models in a decentralized manner without sharing users' private data [19].

Jiachun et al. introduced ADDETECTOR, a privacy-preserving system for Alzheimer's disease (AD) detection using IoT devices and security mechanisms. ADDETECTOR collects user audio data from smart home IoT devices and utilizes novel linguistic features to enhance detection accuracy. Employing FL and differential privacy mechanisms, ADDETECTOR ensures user data privacy across three layers: user, client, and cloud. Evaluation of 1010 AD detection trials demonstrates ADDETECTOR's high accuracy of 81.9% and low time overhead of 0.7 s, showcasing its effectiveness in preserving privacy while achieving accurate AD detection in smart home environments [20].

Dasara Dharami Reddy et al. reviewed recent advancements in FL for healthcare applications. The survey outlines FL fundamentals, motivations, and applications across various healthcare domains. It highlights lessons learned, challenges, and future directions, providing insights for prospective researchers interested in FL for healthcare informatics [21].

Bjarne et al. presented a systematic literature review on FL in the medical domain. FL is explored as a solution for combining data from different sites while preserving privacy. The review covers FL concepts, research, architectures, challenges, attacks, defenses, and ethical considerations, providing an extensive overview of FL's applicability for confidential healthcare datasets [22].

A thorough analysis of FL for privacy preservation in smart healthcare systems was presented by Mansoor et al., with an emphasis on privacy-related concerns in the Internet of Medical Things (IoMT). The review highlights the critical role that FL plays in protecting privacy in IoMT networks and presents sophisticated FL architectures that use digital twins, generative adversarial networks (GANs), and deep reinforcement learning (DRL) to identify privacy concerns. Mansoor et al. reviewed previous surveys in FL in smart healthcare and tallied existing work based on privacy hazards and preservation approaches utilizing FL in IoMT networks after conducting a thorough evaluation of 109 research and survey articles. Additionally, the paper highlighted some FL applications in smart IoMT healthcare systems and explored the practical potential for FL in IoMT. Finally, the poll highlights the need for more study and development in this area by talking about open research concerns and challenges in utilizing FL in future smart healthcare systems [23].

The privacy issues surrounding centralized global aggregation in FL systems were discussed by Mahmuda et al., especially in relation to smart healthcare systems. Acknowledging the risks associated with inference attacks, free-riding, and direct manipulation by central servers, the paper presented Edge Intelligence, a three-fold Federated Edge Aggregator (FEA) framework intended to protect privacy attacks against Smart Healthcare Systems at the edge. To strike a compromise between privacy preservation and model performance, Edge Intelligence leverages FL and an iteration-based Conventional Neural Network (CNN) model with artificial noise functions. The authors also proved a theoretical bound on Edge Intelligence's convergence on the loss function of the trained FL model. Mahmuda et al. proved the effectiveness of their architecture by evaluating it on well-known datasets such as the COVID-19 chest x-ray, MNIST, CIFAR10, STL10, and 90% accuracy when compared to baseline methods [24].

With an emphasis on improving privacy protection, Zhang et al. incorporated FL into the deep learning of medical models in Internet of Things (IoT)-enabled healthcare systems. The study additionally protected local models by utilizing cryptographic primitives like masks and homomorphic encryption to stop attackers from deducing confidential medical information through a variety of techniques such as model reconstruction or inversion. The authors also suggested a dropout-tolerable FL method to guarantee the process's continuation even in the event that the number of online clients stays above a certain threshold. They proved through security research how well their suggested plan would protect data privacy. The study also included theoretical evaluations of the scheme's communication and computation costs. Skin lesion classification using the HAM10000 medical dataset was done to show how their method is actually used in healthcare. The scheme's usefulness was demonstrated by experimental results, which showed that it outperformed other schemes in terms of privacy preservation and produced promising results in the classification of skin lesions [25].

The difficulties FL had in satisfying the strict specifications of healthcare systems built on numerous Internet of Medical Things (IoMT) were discussed by Aouedi et al. Even while FL systems are designed to prevent participants from sharing data, there are still security issues that present vulnerabilities from several directions. The study included a thorough analysis of current strategies and issues surrounding the nascent application of FL in healthcare. In order to ensure that FL is successfully integrated into healthcare systems, it emphasized the necessity of further work addressing security and privacy issues in federated healthcare [26].

Suraj et al. used electronic health record data from two academic medical centers on a Microsoft Azure Cloud Data bricks platform to investigate multiple FL implementations utilizing both simulated and real-world settings. An impressive 3% gain in performance was shown by the cyclically trained Artificial Neural Network (ANN), which was a substantial improvement over the majority of the efforts ($P < .05$). After FL procedures, Logistic regression (LR) models did not significantly improve; however, single weight neural network models showed improvement in certain circumstances. The authors observed that depending on the machine learning model and the FL setup, different processes led to different performance improvements [27]. They also verified that the sequence in which the institutions were trained affected the overall improvement in performance. Let us check Table 9.1, where some significant existing research works are reviewed very carefully:

TABLE 9.1
Analysis of Existing Work on FL-Based Healthcare System

Serial No	Reference	Their Contribution	Citation
1	[3]	Their approach introduces a Blockchain and FL-enabled Secure Architecture for Privacy-Preserving in Smart Healthcare, enabling the utilization of well-trained machine learning models without compromising sensitive data to the cloud. By leveraging FL, their framework facilitates scalable machine learning applications in the healthcare industry while ensuring data privacy and security.	The complexity of implementing blockchain and FL technologies in healthcare systems, potential challenges in interoperability with existing infrastructure, and the need for further validation in real-world healthcare settings to assess scalability, efficiency, and usability.
2	[4]	Introducing a Deep FL framework for decentralized healthcare systems, incorporating an automated training data acquisition process, and demonstrating its application in detecting skin diseases. Transfer Learning is employed to mitigate limited healthcare data availability, resulting in a significant improvement in model performance, with FL increasing the Area Under the Curve of the centralized learning model up to 0.97.	Potential impact on quality of service due to increased model conversion time in the FL system. Additionally, the framework's effectiveness and scalability in real-world healthcare settings with diverse datasets and operational constraints may require further validation and investigation.
3	[5]	Proposing an FL-based Privacy-Preserving System, ADDETECTOR, for Alzheimer's disease detection using IoT devices, ensuring user data privacy and model confidentiality.	Limited discussion on potential challenges or drawbacks of implementing ADDETECTOR in real-world healthcare settings, requiring further investigation into scalability and usability.
4	[9]	Introducing a three-fold Federated Edge Aggregator framework to protect Smart Healthcare Systems against privacy attacks, leveraging FL and Edge Intelligence.	Absence of empirical validation or real-world implementation of the proposed framework, necessitating further validation and testing in healthcare settings.
5	[10]	Introducing a Homomorphic Encryption-based Privacy-Preserving FL approach for skin lesion classification, safeguarding sensitive medical data while achieving effective model training.	Lack of detailed discussion on the scalability and computational overhead of implementing the proposed FL scheme in real-world healthcare settings, warranting further investigation and validation.
10	[12]	They used electronic health record data from two academic medical centers on a Microsoft Azure Cloud Data bricks platform to investigate multiple FL implementations utilizing both simulated and real-world settings.	Limited addressing of class distribution inequality and potential overfitting in FL due to varying dataset sizes across clients.

9.3 IMPLEMENTATION OF FL IN HEALTHCARE SYSTEM

FL is a novel machine learning idea and methodology. It has enormous potential to change the healthcare sector. Professionals in healthcare may also benefit more from it. FL doesn't want to take the role of medical experts. Its goal is to free them up to focus on providing better care for patients [28]. The ability to access vast volumes of data for analysis and model training is essential for the innovation of healthcare in the future. FL is a potential approach to enhance patient care and progress the healthcare industry since it is an effective tool for gaining access to data without compromising data privacy. Implementing a FL project effectively and having a beneficial influence on the healthcare business may be achieved by following these steps and taking the required safeguards to preserve patient privacy [29].

Figure 9.2 shows an existing architecture of FL-based healthcare system. In this system hospitals have an edge server in which data computations are done locally. These edge servers are connected with the global server. The pandemic crisis has caused us to go through a great deal of upheaval in the last year. During this period, it was clear that the healthcare sector lacked resources. To help them treat patients more effectively, healthcare workers require dependable technology. Large and diverse data sets would be necessary for training an algorithm for clinical applications. Strict laws such as HIPPA make it more difficult to provide important information [30].

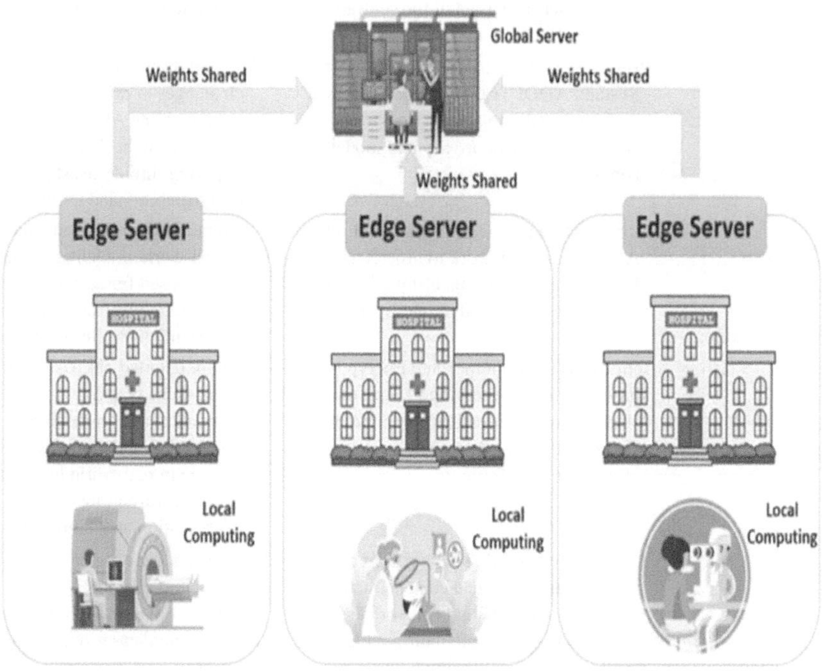

FIGURE 9.2 Existing architecture of FL-based smart healthcare system.

FL was introduced in this situation. On their internal data pool, participating institutions utilize the same algorithm for training. These algorithms are taught and might be useful to these institutions. Their ability to access and circumvent different restrictions would be facilitated. Additionally, it provides access to a data set that might be used for learning.

Due to its ability to generate sizable patient data databases that researchers may utilize for their studies without requiring direct access to any person's personal health information, FL has been applied in medical research. It is challenging to expand machine learning systems in the healthcare sector internationally due to the sensitive nature of the data and its restricted access due to privacy concerns [31]. FL allows training models to use patient and medical institution data while maintaining secure access to the original data source. It allows the models to safely learn from additional datasets and can facilitate collaboration between specific universities. FL can also provide access to cutting-edge AI technology for smaller rural hospitals and enable physicians to learn more about patients or diseases from demographic regions beyond local institutions [32].

Due to its capacity to give healthcare organizations superior insights into the patients' medical histories and treatment plans, FL has been increasingly popular in recent years. However, for FL applications to succeed, they must first be properly deployed by healthcare professionals who are aware of the advantages these systems may offer to their organizations [33].

9.3.1 How to Use FL in Healthcare Organization

There are three basic steps that require following for implementing FL in healthcare organizations. These are as follows:

1. **Establish the objective**

Ascertain the necessary goals for FL. Does the model's accuracy need to be increased? Is a decrease in the quantity of training instances required to train the particular model anticipated? Is it necessary to enhance the model's performance over time?

2. **Pick a suitable FL framework**

There are several FL frameworks available, and each has advantages and disadvantages of its own. The decision will be based on the type of model being built and the type of data set that is available. If a certain framework's compatibility with a particular healthcare provider's infrastructure is crucial for the task at hand, it must also be taken into account.

3. **Set up the FL framework such that it may be used in medical environments**

It is advisable to conduct practical tests of the FL framework prior to implementing it in production, particularly if it is being utilized for making choices on patient care! This can minimize any possible problems during deployment caused by unforeseen events or other variables outside of regular operation settings and assist in guaranteeing that it functions as intended in real-world scenarios [34].

9.3.2 Healthcare Applications Using FL

In the field of healthcare, FL has several applications.

Disease Prediction: FL models may be used by hospitals to estimate the likelihood that a certain patient will become unwell or contract an infection after being admitted. With each patient's individual needs in mind, this might assist physicians decide how to treat them most effectively. Moreover, other healthcare-related organizations (like insurance firms) may utilize these forecasts to offer better coverage alternatives to hospital patients who most need them [35].

Disease Understanding: A further way that FL might improve healthcare is by enhancing our genetic understanding of conditions like cancer and Alzheimer's disease. By using FL models on DNA samples obtained from thousands of individuals afflicted with various illnesses, scientists can discover the elements that most significantly contribute to the development of these diseases [36].

Research: Since FL eliminates the requirement for direct access to any individual's private health information (PHI), researchers are able to generate sizable patient data datasets for use in their investigations. Additionally, it enables researchers to access resources (such as genetic information about patients) through the databases of other researchers, even if they may not otherwise have access to particular medical records or resources. In doing so, privacy violations are avoided while providing researchers with sufficient data and insights [37].

Precision Medicine: Large-scale precision medicine research is one of the most potential uses of FL, which is a rapidly expanding topic in healthcare. Experimental research that employs personal data to create a customized treatment plan for each patient is known as a precision medicine study. Targeting therapy according to each person's distinct genetic profile and medical background is the aim of these investigations. FL has applications in both clinical and pharmaceutical settings, including the prediction of drug reactions and other treatment results, as well as the advancement of diseases and patient outcomes. Researchers may be able to find novel ways to treat diseases like cancer and Alzheimer's disease at the molecular level with the use of FL, which might result in more individualized therapies for people suffering from these illnesses [38].

9.4 FL AND BLOCKCHAIN INTEGRATION FOR SMART HEALTHCARE SYSTEM

Secure, decentralized, and privacy-preserving solutions for healthcare data management and AI model training may be found in a smart healthcare system built on FL and blockchain. The quality of AI models may be enhanced while addressing privacy and security issues in the healthcare industry by integrating FL for privacy-preserving model training with blockchain for safe data management.

Figure 9.3 illustrates the framework of FL and blockchain-based smart healthcare system. In this framework, a multiple number of hospitals can be connected in a single healthcare system where each hospital has a separate electronic health record (EHR). Local training model is used to compute or process data comes from the each EHR of the hospital in the healthcare system. The entire local training model is aggregated through the model aggregation module. After the model aggregation procedure, all the computed health-related information is transferred to the central server. With the help of blockchain technology this information is stored and shared back to the local training model. Benefits of the suggested framework mainly include:

Data Privacy: Because patient information is not directly exchanged, it is kept secure and confidential.

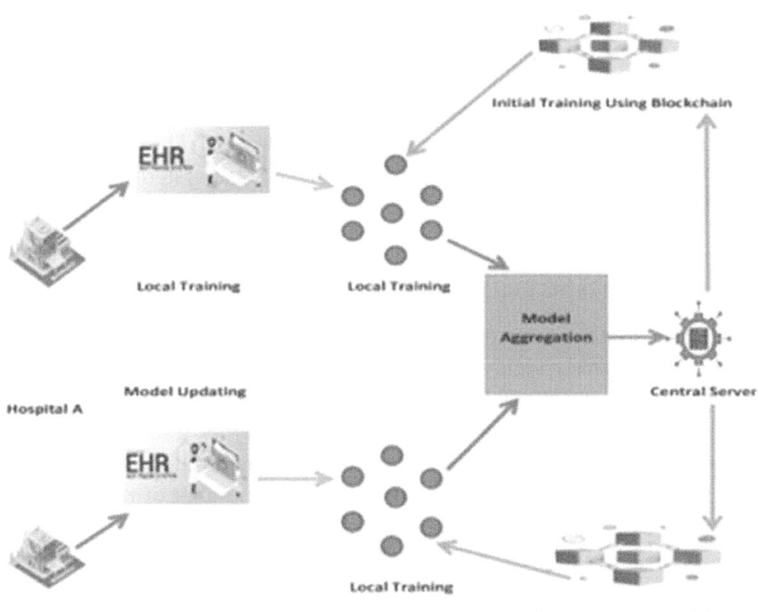

FIGURE 9.3 Framework of FL and blockchain-based smart healthcare system.

Model Efficiency: Based on local data, local training models get better over time, producing more accurate and customized models.

Security: Data integrity and access control are guaranteed by blockchain technology, and data is safeguarded throughout model aggregation through FL.

9.5 ISSUES AND CHALLENGES OF FL IN HEALTHCARE SYSTEMS

Organizations should be careful to select a federated machine learning platform that leverages effective communication to tackle this difficulty, regardless of their computational resources, as FL demands a lot of compute and bandwidth availability. Without requiring data to be moved outside of the custodian's environment, the appropriate platform can handle heavy workloads, numerous calculations, and different types of collaborations [39]. Inadequate platforms are unable to accomplish this because they are difficult to incorporate into the required workflows. Rather, if they do have the chance to train on distinct datasets, they sit beside them, necessitating the lengthy and expensive rewriting of models. These platforms are not designed for smooth, scalable FL because they do not comply with regulatory requirements, integrate with current workflows and stacks, or lessen the technical burden [40].

FL in healthcare has a number of difficulties. The scarcity of data, particularly in industries like healthcare where enormous volumes of data are needed to train AI models, is a significant obstacle. Concerns about privacy that arise with exchanging medical records provide another difficulty [41]. The majority of the time, healthcare data is dispersed among several sources, and these sources frequently hesitate to exchange data because of privacy concerns. In FL-enabled IoT systems for health monitoring, patient data confidentiality may be threatened by malicious ML algorithm coordination. Furthermore, there are new issues with interoperability, data management, and security as a result of the integration of emerging technologies like blockchain, AI, cloud edge computing, and the Internet of Things in the healthcare industry. The successful integration of FL in healthcare systems depends on overcoming these obstacles [42].

9.5.1 Privacy and Security Concerns

Multiple data sets that were previously hidden behind sensitivity issues and legal limitations would surely yield better insights than one could ever hope for, just like two heads are always preferable to one. The issue is that businesses are reluctant to share sensitive data, intellectual property (IP), and other trade secrets when they are trying to enhance their own data sets with partners or internally across multiple locations. In addition to challenging implicit trust, a number of regional regulatory hurdles must be cleared for this to occur [43].

Although FL doesn't provide a complete solution to privacy and security issues, it can alleviate some of the anxiety by eliminating the need to transfer data. Despite this, further measures must be put in place to make sure that intellectual property and other private information are protected from disclosure when the user receives their findings [44].

In order to do this, companies should use a federated infrastructure to provide a modular approach to privacy-enhancing technologies (PETs) inside their FL strategy. By combining a federated and secure infrastructure with identity and access control at the computational level, these companies can guarantee optimal data insights while minimizing privacy issues. Due to the large, unique, and multifaceted nature of security, no two organizations will have precisely the same security-related challenges. As a result, there are several more factors to take into account. In order to facilitate meaningful data cooperation, each organization must supplement FL with the appropriate IT security procedures and technologies [45].

Because medical data is sensitive, FL in the healthcare industry has special privacy and security problems. Among the principal problems and obstacles are:

Data privacy: FL uses several decentralized data sources to train models without sharing raw data. But it's imperative to make sure that patient data is confidential and anonymous while the model is being trained [46].

Information security: It is crucial to safeguard information from hostile assaults, illegal access, and data breaches. It is imperative to use secure communication protocols and encryption [47].

Model Poisoning: The integrity and functionality of the model may be jeopardized by malicious players attempting to manipulate it by inserting poisoned data.

Fairness and Bias: Data biases can provide discriminating or unjust results. Biases must be taken into consideration and mitigated via FL algorithms.

Regulatory Compliance: Strict laws (such as GDPR in the EU and HIPAA in the US) apply to healthcare data. These rules must be followed by FL systems.

Heterogeneity of Participation: The amount, quality, and format of data sources in the healthcare industry vary. FL has to take this variation into consideration.

Resources Restrictions: The computing power of devices used in healthcare environments may be constrained. Lightweight and efficient algorithms are essential for FL.

9.6 ADDRESSING PRIVACY AND SECURITY-RELATED ISSUES AND CHALLENGES OF FL-BASED HEALTHCARE SYSTEM

A multifaceted strategy involving organizational, technological, and legal safeguards is needed to address privacy and security issues in FL-based healthcare systems. Here are a few tactics:

Encrypt data: Data should be encrypted while it's in transit and at rest by using methods like homomorphic encryption. As a result, even when data is utilized for model training, confidentiality is guaranteed [48].

Differential Privacy: To ensure that specific data points cannot be recognized, separate the data before sharing it by adding noise [49].

Secure Aggregation: Ensuring privacy by aggregating model updates from various devices through the use of safe aggregation algorithms that conceal individual contributions [50].

Data Minimization: Send just the information required for model updates to devices to reduce the quantity of data that is exchanged between them.

Participant Authentication: Put robust authentication in place to guarantee that FL can only be accessed by authorized devices.

Model Validation: Validate models to make sure they don't have biases and don't go against security or privacy laws.

Regulatory Compliance: Adhere to applicable laws and regulations (such as GDPR and HIPAA) by putting in place the appropriate security and privacy safeguards.

Secure Communication: To safeguard data while it's being sent between devices and servers use secure communication protocols (like TLS) [51].

Data anonymization: To safeguard patient identities during FL, anonymize data before sharing it.

Education and Awareness: Spread the word among patients, data scientists, and healthcare professionals about the value of security and privacy in FL.

9.7 ALGORITHMS TO ADDRESS THE PRIVACY AND SECURITY-RELATED ISSUES AND CHALLENGES OF FL-BASED HEALTHCARE SYSTEM

Healthcare systems can use a variety of FL methods to handle privacy and security concerns related to FL. One well-known technique that is frequently utilized because of its efficiency and simplicity is Federated Averaging (FedAvg). In order to ensure that no client's data is exposed throughout the training process, FedAvg operates by averaging the model updates from several customers. While maintaining data privacy, the FedAvg algorithm facilitates cooperative model training across several clients. Clients only provide the server updates about their models—not raw data—and the server compiles these updates into a global model [52]. A more reliable and broadly applicable global model is produced by averaging the local models, which lessens the influence of client-specific noise and biases.

FedAvg is a useful technique; but, by including noise into the model updates prior to aggregation, alternative algorithms, such FL with Differential Privacy (DP-FedAvg), can improve privacy even further while maintaining the confidentiality of individual client contributions [53]. When it comes to extremely sensitive healthcare applications that demand stringent privacy assurances, DP-FedAvg is especially helpful. Figure 9.4 shows the basic steps of FedAvg algorithms to be followed carefully to ensure the highest level of privacy and security in any commercial applications including healthcare sector.

FIGURE 9.4 Block diagram of FedAvg algorithm.

The following are some ways that FedAvg can handle security and privacy concerns in healthcare systems:

Confidentiality Conservation: FedAvg protects user privacy by only transmitting updated model data—not raw data to the central server for aggregation. This keeps private patient information safe.

Data security: The danger of data breaches and unauthorized access is decreased because raw data is kept on the client side. Transmitting data is encrypted to increase security even more.

Fairness and Bias Reduction: FedAvg aggregates updates from a wide range of customers, therefore reflecting a more complete population, which can assist assure fairness and minimize bias in healthcare AI models.

Regulatory Compliance: FedAvg's commitment to protecting user privacy complies with laws like GDPR and HIPAA, which demand that patient data be kept secure.

Effective usage of Resources: FedAvg minimizes the requirement for data centralization and its related expenses by enabling the usage of local data on client devices.

In a smart healthcare system, there are several crucial phases to implementing the Differential Privacy Federated Averaging (DP FedAvg) algorithm. To guarantee adherence to privacy laws, healthcare data is first gathered and preprocessed from several sources, including hospitals and wearable technology. Next, each training round's participants are chosen from among the healthcare practitioners. To protect the privacy of patient data, differential privacy noise is then created and added to the gradients computed during training. Upon initialization, the global model is forwarded to the designated clients, who utilize their own data to train it locally. A new global model is created by aggregating the updated models using FedAvg after training. Until convergence, several rounds of this technique are performed. After training, the global model is used to forecast fresh data, and ongoing maintenance and monitoring procedures assure the model's continued privacy and performance.

A version of the Federated Averaging (FedAvg) algorithm that uses differential privacy measures is called Differential Privacy Federated Averaging (DP FedAvg). It provides further advantages and benefits in relation to intelligent healthcare systems: Improved confidentiality protection, enhanced trust and clearness, conservation of data duality, robustness against privacy threat, and simplification of cross institutional collaboration etc. To summarize, DP FedAvg plays a crucial role in smart healthcare systems by safeguarding patient privacy, guaranteeing adherence to regulations, cultivating confidence, and promoting teamwork, all while preserving the effectiveness of machine learning models.

9.8 CONCLUSION

The delicate nature of medical data makes privacy and security crucial concerns in FL-based healthcare systems, to sum up. Robust model aggregation approaches, improved privacy-preserving procedures, and secure communication protocols are just a few of the many strategies needed to identify and overcome these difficulties. In order to secure patient data during model training and aggregation, researchers can create enhanced privacy strategies including homomorphic encryption and differential privacy. To manage diverse data sources and reduce the danger of model poisoning assaults, robust model aggregation strategies are required.

Incorporating blockchain technology into FL systems may also improve data security, integrity, and traceability while guaranteeing adherence to laws like GDPR and HIPAA." While preserving confidentiality and privacy, dynamic and adaptive FL algorithms can adapt to shifting data distributions and network circumstances. To verify the efficiency and scalability of security and privacy measures in FL-based healthcare systems, real-world deployment studies are crucial. A comprehensive approach to tackling these difficulties requires interdisciplinary collaboration involving computer scientists, healthcare practitioners, legislators, and ethicists. In general, researchers may progress the area and guarantee that patient data is secured in AI-driven healthcare solutions by tackling privacy and security challenges in FL-based healthcare systems.

9.9 FUTURE WORK

Data In order to solve new problems and advance the field, future research on privacy and security issues in FL-based healthcare systems is essential. Some important topics requiring more study include:

Improved Privacy Methods: Creation of more sophisticated privacy-preserving methods to safeguard private medical data during model training and aggregation, such as secure multiparty computing (MPC), and differential privacy.

Protected Communication Standard: Formation and improvement of secure communication methods to ensure data integrity and secrecy when sending aggregated models and model changes between clients and servers.

Adaptive and Vibrant FL: Generating dynamic and adaptable FL algorithms that can adapt to shifting client involvement, data distributions, and network circumstances without compromising security and privacy is the goal of this work.

User Awareness: Research the effects of user education and awareness initiatives on security and privacy procedures pertaining to patient data in FL-based healthcare systems.

Interdisciplinary Association: To tackle the privacy and security issues in FL-based healthcare systems, it is recommended to foster multidisciplinary collaboration among computer scientists, healthcare practitioners, politicians, and ethicists.

By tackling these research topics, we can further FL in healthcare and ensure that privacy and security are always top priorities in the creation of AI-driven healthcare solutions.

REFERENCES

1. Li, L., Fan, Y., Tse, M. and Lin, K.Y. 2020. A review of applications in federated learning. *Computers & Industrial Engineering*, *149*, p. 106854.
2. Aledhari, M., Razzak, R., Parizi, R.M. and Saeed, F. 2020. Federated learning: A survey on enabling technologies, protocols, and applications. *IEEE Access*, *8*, pp. 140699–140725.
3. Rahman, K.J., Ahmed, F., Akhter, N., Hasan, M., Amin, R., Aziz, K.E., Islam, A.M., Mukta, M.S.H. and Islam, A.N. 2021. Challenges, applications and design aspects of federated learning: A survey. *IEEE Access*, *9*, pp. 124682–124700.
4. Zhang, C., Xie, Y., Bai, H., Yu, B., Li, W. and Gao, Y. 2021. A survey on federated learning. *Knowledge-Based Systems*, *216*, p. 106775.
5. Zheng, Z., Zhou, Y., Sun, Y., Wang, Z., Liu, B. and Li, K. 2022. Applications of federated learning in smart cities: recent advances, taxonomy, and open challenges. *Connection Science*, *34*(1), pp. 1–28.
6. Banabilah, S., Aloqaily, M., Alsayed, E., Malik, N. and Jararweh, Y. 2022. Federated learning review: Fundamentals, enabling technologies, and future applications. *Information Processing & Management*, *59*(6), p. 103061.
7. Vahidian, S., Morafah, M., Chen, C., Shah, M., & Lin, B. (2023). Rethinking data heterogeneity in federated learning: Introducing a new notion and standard benchmarks. *IEEE Transactions on Artificial Intelligence*, *5*(3), 1386–1397.
8. Yin, F., Lin, Z., Kong, Q., Xu, Y., Li, D., Theodoridis, S. and Cui, S.R. 2020. FedLoc: Federated learning framework for data-driven cooperative localization and location data processing. *IEEE Open Journal of Signal Processing*, *1*, pp. 187–215.

9. Kairouz, P., McMahan, H.B., Avent, B., Bellet, A., Bennis, M., Bhagoji, A.N., Bonawitz, K., Charles, Z., Cormode, G., Cummings, R. and D'Oliveira, R.G. 2021. Advances and open problems in federated learning. *Foundations and trends® in machine learning, 14*(1–2), pp. 1–210.
10. Li, Q., Wen, Z., Wu, Z., Hu, S., Wang, N., Li, Y., Liu, X. and He, B. 2021. A survey on federated learning systems: Vision, hype and reality for data privacy and protection. *IEEE Transactions on Knowledge and Data Engineering, 35*(4), pp. 3347–3366.
11. Savazzi, S., Nicoli, M., Bennis, M., Kianoush, S. and Barbieri, L. 2021. Opportunities of federated learning in connected, cooperative, and automated industrial systems. *IEEE Communications Magazine, 59*(2), pp. 16–21.
12. Naik, D. and Naik, N. 2023, September. An introduction to federated learning: working, types, benefits and limitations. In *UK Workshop on Computational Intelligence* (pp. 3–17). Cham: Springer Nature Switzerland.
13. Shaheen, M., Farooq, M.S., Umer, T. and Kim, B.S. 2022. Applications of federated learning; Taxonomy, challenges, and research trends. *Electronics, 11*(4), p. 670.
14. Mansour, Y., Mohri, M., Ro, J. and Suresh, A.T. 2020. Three approaches for personalization with applications to federated learning. *arXiv preprint arXiv:2002.10619.*
15. Zhang, C., Xie, Y., Bai, H., Yu, B., Li, W. and Gao, Y. 2021. A survey on federated learning. *Knowledge-Based Systems, 216*, p. 106775.
16. Thummisetti, B.S.P. and Atluri, H. 2024. Advancing healthcare informatics for empowering privacy and security through federated learning paradigms. *International Journal of Sustainable Development in Computing Science, 1*(1), pp. 1–16.
17. Nguyen, D.C., Quoc-Viet, P., Pathirana, P.N., Ming, D., Seneviratne, A., Zihuai, L.I.N., Dobre, O. and Won-Joo, H. 2023. Federated learning for smart healthcare: a survey. *ACM Computing Surveys, 2023*, pp. 1937-01.
18. Singh, S., Rathore, S., Alfarraj, O., Tolba, A. and Yoon, B. 2022. A framework for privacy-preservation of IoT healthcare data using Federated Learning and blockchain technology. *Future Generation Computer Systems, 129*, pp. 380–388.
19. Elayan, H., Aloqaily, M. and Guizani, M. 2021. Deep federated learning for IoT-based decentralized healthcare systems. In *2021 International Wireless Communications and Mobile Computing (IWCMC)*, pp. 105–109. IEEE.
20. Jiachun, Y. M., Ma, L., Du, S., Zhu, H., Pei, Q. and Shen, X. 2021. A federated learning based privacy-preserving smart healthcare system. *IEEE Transactions on Industrial Informatics, 18*(3), pp. 2021–2031.
21. Reddy, K. D. and Gadekallu, T.R. 2023. A comprehensive survey on federated learning techniques for healthcare informatics. *Computational Intelligence and Neuroscience, 2023(1)*, p. 8393990.
22. Pfitzner, B., Steckhan, N. and Arnrich, B. 2021. Federated learning in a medical context: a systematic literature review. *ACM Transactions on Internet Technology (TOIT), 21*(2), pp. 1–31.
23. Ali, M., Naeem, F., Tariq, M. and Kaddoum, G. 2022. Federated learning for privacy preservation in smart healthcare systems: A comprehensive survey. *IEEE Journal of Biomedical and Health Informatics, 27*(2), pp. 778–789.
24. Akter, M., Moustafa, N., Lynar, T. and Razzak, I. 2022. Edge intelligence: Federated learning-based privacy protection framework for smart healthcare systems. *IEEE Journal of Biomedical and Health Informatics, 26*(12), pp. 5805–5816.
25. Zhang, L., Xu, J., Vijayakumar, P., Sharma, P.K. and Ghosh, U. 2022. Homomorphic encryption-based privacy-preserving federated learning in iot-enabled healthcare system. *IEEE Transactions on Network Science and Engineering, 10*(5), pp. 2864–2880.

26. Aouedi, O., Sacco, A., Piamrat, K. and Marchetto, G. 2022. Handling privacy-sensitive medical data with federated learning: challenges and future directions. *IEEE Journal of Biomedical and Health Informatics*, *27*(2), pp. 790–803.
27. Rajendran, S., Obeid, J.S., Binol, H., Foley, K., Zhang, W., Austin, P., Brakefield, J., Gurcan, M.N. Topaloglu, U. 2011. Cloud-based federated learning implementation across medical centers. *JCO Clinical Cancer Informatics*, *5*(2021), pp. 1–11.
28. Zellinger, W., Wieser, V., Kumar, M., Brunner, D., Shepeleva, N., Gálvez, R., Langer, J., Fischer, L. and Moser, B. 2021. Beyond federated learning: On confidentiality-critical machine learning applications in industry. *Procedia Computer Science*, *180*, pp. 734–743.
29. Nguyen, D.C., Ding, M., Pathirana, P.N., Seneviratne, A., Li, J. and Poor, H.V. 2021. Federated learning for internet of things: A comprehensive survey. *IEEE Communications Surveys & Tutorials*, *23*(3), pp. 1622–1658.
30. Reddy, K.D. and Gadekallu, T.R. 2023. A comprehensive survey on federated learning techniques for healthcare informatics. *Computational Intelligence and Neuroscience*, *2023(1)*, p. 8393990.
31. Yoo, J.H., Jeong, H., Lee, J. and Chung, T.M. 2021. Federated learning: Issues in medical application. In *Future Data and Security Engineering: 8th International Conference, FDSE 2021, Virtual Event, November 24–26, 2021, Proceedings 8* (pp. 3–22). Springer International Publishing.
32. Khan, L.U., Saad, W., Han, Z., Hossain, E. and Hong, C.S. 2021. Federated learning for internet of things: Recent advances, taxonomy, and open challenges. *IEEE Communications Surveys & Tutorials*, *23*(3), pp. 1759–1799.
33. Singh, P., Singh, M.K., Singh, R. and Singh, N. 2022. Federated learning: Challenges, methods, and future directions. In *Federated Learning for IoT Applications* (pp. 199–214). Cham: Springer International Publishing.
34. Xu, J., Glicksberg, B.S., Su, C., Walker, P., Bian, J. and Wang, F. 2021. Federated learning for healthcare informatics. *Journal of Healthcare Informatics Research*, *5*, pp. 1–19.
35. Bebortta, S., Tripathy, S.S., Basheer, S. and Chowdhary, C.L. 2023. FedEHR: a federated learning approach towards the prediction of heart diseases in IoT-based electronic health records. *Diagnostics*, *13*(20), p. 3166.
36. Moshawrab, M., Adda, M., Bouzouane, A., Ibrahim, H. and Raad, A. 2023. Reviewing federated machine learning and its use in diseases prediction. *Sensors*, *23*(4), p. 2112.
37. Rieke, N., Hancox, J., Li, W., Milletari, F., Roth, H.R., Albarqouni, S., Bakas, S., Galtier, M.N., Landman, B.A., Maier-Hein, K. and Ourselin, S. 2020. The future of digital health with federated learning. *NPJ Digital Medicine*, *3*(1), pp. 1–7.
38. Li, S., Cai, T. and Duan, R. 2023. Targeting underrepresented populations in precision medicine: A federated transfer learning approach. *The Annals of Applied Statistics*, *17*(4), pp. 2970–2992.
39. Rahman, A., Hossain, M.S., Muhammad, G., Kundu, D., Debnath, T., Rahman, M., Khan, M.S.I., Tiwari, P. and Band, S.S. 2023. Federated learning-based AI approaches in smart healthcare: concepts, taxonomies, challenges and open issues. *Cluster Computing*, *26*(4), pp. 2271–2311.
40. Rahman, K.J., Ahmed, F., Akhter, N., Hasan, M., Amin, R., Aziz, K.E., Islam, A.M., Mukta, M.S.H. and Islam, A.N. 2021. Challenges, applications and design aspects of federated learning: A survey. *IEEE Access*, *9*, pp. 124682–124700.
41. Bharati, S., Mondal, M., Podder, P. and Prasath, V.B. 2022. Federated learning: Applications, challenges and future directions. *International Journal of Hybrid Intelligent Systems*, *18*(1–2), pp. 19–35.

42. Yoo, J.H., Jeong, H., Lee, J. and Chung, T.M. 2022. Open problems in medical federated learning. *International Journal of Web Information Systems, 18*(2/3), pp. 77–99.
43. Al-Huthaifi, R., Li, T., Huang, W., Gu, J. and Li, C. 2023. Federated learning in smart cities: Privacy and security survey. *Information Sciences, 632*, pp. 833–857.
44. Wang, W., Li, X., Qiu, X., Zhang, X., Brusic, V. and Zhao, J. 2023. A privacy preserving framework for federated learning in smart healthcare systems. *Information Processing & Management, 60*(1), p. 103167.
45. Ma, C., Li, J., Ding, M., Yang, H.H., Shu, F., Quek, T.Q. and Poor, H.V. 2020. On safeguarding privacy and security in the framework of federated learning. *IEEE network, 34*(4), pp. 242–248.
46. Mothukuri, V., Parizi, R.M., Pouriyeh, S., Huang, Y., Dehghantanha, A. and Srivastava, G. 2021. A survey on security and privacy of federated learning. *Future Generation Computer Systems, 115*, pp. 619–640.
47. Akter, M., Moustafa, N., Lynar, T. and Razzak, I. 2022. Edge intelligence: Federated learning-based privacy protection framework for smart healthcare systems. *IEEE Journal of Biomedical and Health Informatics, 26*(12), pp. 5805–5816.
48. Zhang, L., Xu, J., Vijayakumar, P., Sharma, P.K. and Ghosh, U. 2022. Homomorphic encryption-based privacy-preserving federated learning in iot-enabled healthcare system. *IEEE Transactions on Network Science and Engineering, 10*(5), pp. 2864–2880.
49. Wei, K., Li, J., Ding, M., Ma, C., Yang, H.H., Farokhi, F., Jin, S., Quek, T.Q. and Poor, H.V. 2020. Federated learning with differential privacy: Algorithms and performance analysis. *IEEE transactions on information forensics and security, 15*, pp. 3454–3469.
50. Beguier, C. and Tramel, E.W. 2020. Safer: Sparse secure aggregation for federated learning. *arXiv preprint arXiv:2007.14861.*
51. Rehman, A., Abbas, S., Khan, M.A., Ghazal, T.M., Adnan, K.M. and Mosavi, A. 2022. A secure healthcare 5.0 system based on blockchain technology entangled with federated learning technique. *Computers in Biology and Medicine, 150*, p. 106019.
52. Sannara, E.K., Portet, F., Lalanda, P. and German, V.E.G.A. 2021, March. A federated learning aggregation algorithm for pervasive computing: Evaluation and comparison. In *2021 IEEE International Conference on Pervasive Computing and Communications (PerCom)* (pp. 1–10). IEEE.
53. Ay, Ş., Ekinci, E. and Garip, Z. 2024. A brain tumour classification on the magnetic resonance images using convolutional neural network based privacy-preserving federated learning. *International Journal of Imaging Systems and Technology, 34*(1), p.e23018.

10 Privacy-Preserving Federated Transfer Learning for Brain Tumor Multi-Classification in AI-Driven Healthcare

Swetha Ghanta, Ashok Kumar Pradhan, Prasanthi Boyapati, Subhra Prosun Paul, and Laavanya Rachakonda

10.1 INTRODUCTION

10.1.1 Brain Tumor Classification

Brain tumors are distinct in shape and severity and are considered a major factor in public health worldwide. The correct identification and classification of these tumors is essential for making appropriate predictions on outcomes and in choosing the optimal treatment accurately [1]. Advances in medical imaging, including magnetic resonance imaging (MRI), have enabled noninvasive visualization and classification of brain tumors. And MRI can give you high-resolution images of the specific structure and the anomalies in the structure, such as tumors. Nonetheless, interpretation of these images and differential diagnosis of multiple tumor types can be difficult, even for experienced radiologists, due to the complexity associated with brain tumors.

In order to meet these challenges, CAD systems are provided to serve as potential tools to assist clinicians in the analysis of medical images. Deep learning (DL) techniques, in particular the convolutional neural networks (CNNs), have manifested promising capability to facilitate brain tumor classification by learning discriminatory features from medical images. Transfer learning (TL), a branch of DL, has been proposed as an interesting approach able to leverage pre-trained knowledge in order to enhance classification performance even when a large amount of annotated data is not at a glance-minable.

DOI: 10.1201/9781003567585-10

10.1.2 Importance of TL in Medical Imaging

In medical imaging, TL [2] has become a potent approach because it helps to overcome challenges like limited labeled data and the need for high computational resources. In the context of brain tumor classification, TL helps improve the accuracy, robustness, and efficiency of machine learning (ML) models by leveraging knowledge from diverse datasets and domains.

Figure 10.1 illustrates the working of TL, where models are initially trained on the large ImageNet dataset for image classification. To fine-tune for the specific task, the classification layer at the end is detached, and the model is adjusted accordingly.

10.1.3 Motivation for the FL Approach

The healthcare sector has a number of challenges with traditional centralized-based approaches, notwithstanding the benefits of TL. The issue pertains to data privacy due to the substantial volume of health-related information. Legislation such as Europe's GDPR and America's HIPAA imposes stringent regulations on the handling of this data. However, in fact, the centralized collection of patient data for training purposes may pose risks of privacy infringements and potential data breaches. A significant difficulty is the heterogeneity of data; medical pictures can differ substantially between institutions due to variations in acquisition techniques, equipment, and patient demographics. No one desires APC models that have been solely trained on regular or homogeneous data to underperform across diverse clinical scenarios. Moreover, healthcare datasets are often characterized by class imbalance and distribution shifts, meaning that some diseases are represented by fewer samples or derived from non-uniformly sampled populations. The consequence of this may be models that are biased and have poor generalization to new scenarios or less common cases.

To deal with these problems, FL provides a logical answer by facilitating a cooperative model training across decentralized data sources while maintaining data confidentiality and privacy [3]. The working of FL is demonstrated in Figure 10.2. In

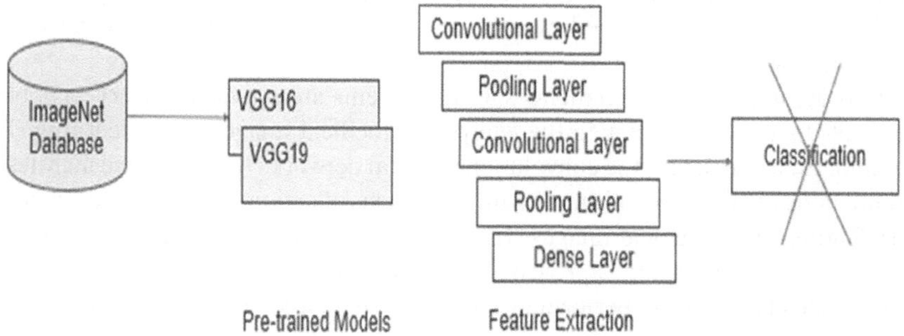

FIGURE 10.1 Working of TL.

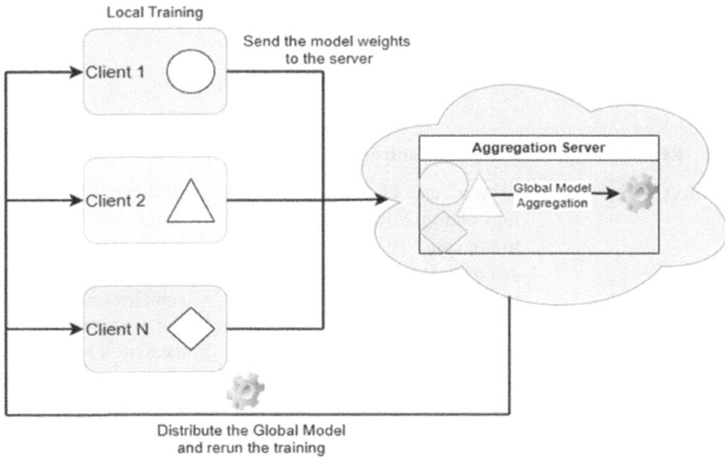

FIGURE 10.2 Working of FL.

FL, ML/DL models are trained locally on data stored on individual institutions (e.g., hospitals, clinics), as opposed to centralizing data on a single server. Only model updates, i.e., the weights, are aggregated and shared with a central server for global model updates. The updated global model is distributed to every client for an additional training cycle. Until convergence or a predetermined number of rounds, this iterative procedure is carried out. FL offers several compelling advantages for medical imaging applications, such as:

- Privacy-Preserving Model Training
- Collaborative Knowledge Sharing
- Robustness to Data Heterogeneity

By combining the principles of FL with TL, the aim is to develop a privacy-preserving and robust framework for multi-class brain tumor classification that leverages the collective knowledge embedded in decentralized data sources while addressing the unique challenges posed by medical imaging data.

10.2 RELATED WORK

There is increasing interest in the possibility of FL to solve privacy, scalability, and efficiency concerns in distributed ML environments, according to a survey of the literature published over the past few years on FL concepts and components in various applications. Table 10.1 summarizes the work and corresponding limitations of significant research papers on the principles and components of FL in various healthcare applications, which are sourced from standard research articles.

The authors Konecny et al. have suggested a new gradient compression method that minimizes the size of gradients sent back and forth between clients and the

TABLE 10.1
Summary of References

SL No.	Reference	Contribution	Limitation
1	Islam et al. [9]	This work uses FL to identify brain tumors from MRI scans, attempting to overcome the problem of centralized data collecting. This work further demonstrates that, in comparison to the conventional DL technique, the FL approach may accomplish privacy-protected tumor classification from MRI scans without significantly sacrificing accuracy.	Regarding the inequality in class distribution among the clients, nothing was done. If the ensemble model has a rather complex neural network design and multiple clients have small datasets, it may significantly overfit, eventually lowering the global performance.
2	Tedeschini et al. [10]	This paper explores the examination of several FL techniques and provides a real-time distributed networking architecture based on the Message Queuing Telemetry Transport (MQTT) protocol. The suggested method is tested in the context of brain tumor segmentation using representative clinical datasets taken from the routine clinical workflow and a modified variation of the popular U-Net model.	The communication bottleneck in the FL process should be carefully addressed when it is implemented over wireless channels. This can be done by using techniques like quantization, compression, and sparsification of model updates. The sizes of the trained models that are commonly seen in medical applications range from 30 to 150 MB for each medical node and learning cycle.
3	Elmas et al. [11]	Federated learning of Generative IMage Priors (FedGIMP) for MRI reconstruction is presented in this paper. FedGIMP uses a two-step process: prior adaptation after the imaging operator is injected, and cross-site learning of a generative MRI prior.	Computational complexity, communication overhead, restricted generalization, and resource constraints are the key limitations of this proposed approach.
4	Zhang et al. [12]	SplitAVG, a novel heterogeneity-aware FL technique, has been developed by the authors of this work to address performance decreases caused by data heterogeneity in FL.	Sensitivity to Heterogeneity, restricted scalability, and dependency on data distribution are the key limitations of this approach.
5	Nazir et al. [13]	The authors of this work have evaluated the uses of FL in medical image analysis using DNNs, emphasized security issues, discussed several attempts to enhance FL model performance, and outlined obstacles and potential future research areas.	Communications, data heterogeneity, data bias, and blockchain implementation are the fundamental challenges of this proposed approach in this paper.

server, leading to quicker convergence and less communication overhead. When compared to conventional FL techniques, the strategy significantly improves communication efficiency, as demonstrated by an evaluation on an actual dataset [4]. A privacy-preserving FL system for medical applications was presented by the authors in a different study. To ensure patient privacy during the model aggregation, a secure aggregation was considered. The framework is used on an electronic health record (EHR) dataset to show how well it defends patient privacy while performing on par with centralized learning methods [5]. Another study discussed how edge devices in Internet of Things (IoT) networks may employ FL. It suggested a low-weight FL method that can be implemented on devices with limited resources, allowing for cooperative model training independent of a centralized server. Based on the experimental results, the suggested method can perform competitively while using less energy and requiring less communication overhead [6].

The use of FL in personalized recommendation systems is examined. It presented a unique architecture for FL that protects user privacy while enabling several clients to cooperatively train a recommendation model. The methodology is assessed on an actual dataset, showcasing its capacity to enhance precision in contrast to traditional centralized approaches [7]. A scalable FL system for large-scale industrial IoT applications is presented in another study. It used an FL architecture that is hierarchical and supported a large number of FL nodes and devices. The scalability and efficiency of the system in training ML models across distributed devices are demonstrated by its application to an industrial sensor data collection [8]. These studies showcase the diverse uses of FL in many business contexts and provide innovative methods to enhance its effectiveness, expandability, and privacy-preserving characteristics.

10.3 FEDERATED LEARNING

To understand the data distribution and make accurate predictions, any ML/DL model requires effective learning. Researchers have identified various types of learning, each with its trade-offs, as illustrated in Figure 10.3. One is local learning, where clients train the models using their local data. This approach is secure, yet the clients' data is limited, and the models may lack generalizability.

The other type is central learning, where clients share their local data with a central cloud server, and the model is trained on this aggregated data. While this

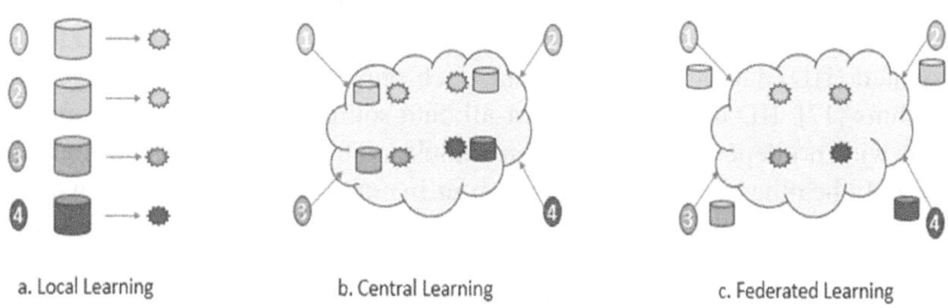

FIGURE 10.3 Different types of learning.

approach addresses the data scarcity and generalizability issues of local learning, it is not secure due to the sharing of sensitive personal data with an untrustworthy third-party cloud server.

To address these issues, a new paradigm called FL is proposed, which allows many parties to work together on creating a shared ML/DL model, eliminating the need for direct sharing of data by employing a decentralized approach for model training.

10.3.1 Definition and Principles of FL

FL is a machine learning paradigm that preserves data privacy by enabling model training across distributed data sources. In FL, model training takes place locally on edge devices, such as smartphones and IoT devices, as opposed to centralizing training data on a single cloud server. Only the model updates are exchanged and are referred to as model weights. This decentralized approach eliminates the need for raw data sharing and protects privacy, making FL particularly well-suited for healthcare applications. The key principles of FL include decentralization, collaborative learning, and privacy preservation.

10.3.2 Relevant FL Algorithms

Some popular FL algorithms used for decentralized model training include Federated Averaging (FedAvg), FL with Secure Aggregation (FedSA), and Differential Privacy (DP). FedAvg is a simple yet effective algorithm for FL, where model updates are averaged to update a global model. It is widely used due to its simplicity [3]. FedSA, on the other hand, enhances privacy in FL by including cryptographic methods such as Homomorphic Encryption (HE) and secure multiparty computation (SMPC) to aggregate model updates without disclosing actual inputs [14]. DP offers strong privacy protection by adding noise, thereby preventing adversaries from obtaining sensitive information about individual data samples [15, 16].

10.4 DATA DISTRIBUTION AND TYPES OF FL

10.4.1 Data Distribution: IID versus Non-IID

Data distribution impacts the performance and reliability of FL, especially in decentralized settings. A key distinction lies between independent and identically distributed (IID) data and non-IID data, each presenting unique challenges for FL algorithms [17]. IID data assumes that all data sources draw from the same distribution, with no dependencies between samples. This uniformity simplifies model training. In healthcare, IID data might occur in controlled environments where data distribution is standard across sites. However, such ideal conditions are rare in real-world scenarios, limiting the practical relevance of IID assumptions in FL.

Non-IID data, on the other hand, reflects the reality of most healthcare environments. Differences in patient populations, disease prevalence, imaging modality, and data collection practices lead to several inconsistencies across institutions. These variations introduce challenges such as uneven model updates, slower convergence, and bias, which can affect generalization to new or underrepresented data. Recognizing and addressing the implications of IID vs. non-IID data is crucial for building generalizable FL models in healthcare.

10.4.2 Types of FL

To address certain challenges in decentralized model training, FL offers several variations like horizontal FL, vertical FL, hierarchical FL, and FTL [18, 19]. Understanding these FL approaches provides insights into their unique characteristics, applications, advantages, and limitations in healthcare environments. Horizontal FL works by combining model updates from separate sources that use the same set of features but contain different samples/individuals. This approach is particularly useful in healthcare, where hospitals can work together on predictive models while keeping patient data private. Sometimes it struggles with changes in data quality or distribution across hospitals, which may affect the model's performance.

Vertical FL is designed for scenarios where different institutions hold complementary data about the same individuals, each with unique types of information, such as one having medical images and another holding EHR. Instead of sharing data, these sources collaborate by training a shared model that combines knowledge from their distinct feature sets. Vertical FL presents technical challenges, such as aligning different data types and coordinating model updates across domains. Hierarchical FL groups participating nodes into clusters or tiers, based on factors like computational power or geographic location. This setup allows for model updates to be aggregated first within clusters and then across them. Its layered design makes it flexible and scalable, but it can also introduce challenges, such as increased communication overhead and difficulty managing differences in data across clusters.

The FTL approach combines the strengths of TL and FL by using pre-trained models' knowledge to enhance the performance of specific tasks [20, 21]. Models trained on general image datasets can be fine-tuned and adapted to specific diseases or imaging categories, leading to faster training and more accurate predictions. FTL boosts generalization, speeds up convergence, and helps models handle diverse data. However, it also comes with issues like model drifting and difficulty transferring knowledge between different domains, requiring careful model selection and adaptation.

10.5 METHODOLOGY

This section discusses the steps taken to train a DL model for brain tumor multiclassification, including data collection, pre-processing, and the FL architecture.

Algorithm 1 Proposed Framework for Privacy-Preserving FTL in Brain Tumor Multi-Classification

Input: Brain MRI images and num_rounds

Output: Tumor Class Label and accuracy values
Begin
Step 1: Load the Brain MRI Images Dataset
Step 2: Resize the images to a standard resolution of 512x512.
Step 3: Split the dataset into training, validation, and testing sub-groups.
Step 4: Develop an FTL-based global model for multi-classification.
Step 5: Set up an FL framework with 3 client nodes.
Step 6: Distribute the data among the clients equally.
Step 7: Initialize and update the local client models with the global model replica.
Step 8: Apply FTL Model - Train and test the model.
Step 9: Obtain the weights from each client node and perform aggregation using FedAvg.
Step 10: Update the global model weights with the aggregated weights.
Step 11: Repeat steps 7 through 10 for num_rounds.
Step 10: Evaluate the model using testing data.
End

10.5.1 Dataset and Pre-processing

We considered the Figshare dataset, which consists of 3064 MRI T1-weighted brain tumor images including meningioma, glioma, and pituitary classes [22]. The sample images can be viewed in Figure 10.4, and the distribution of class labels is depicted in Figure 10.5.

Before model training, the medical imaging data undergoes pre-processing to enhance model performance and generalization. Several preprocessing methods are used before the images are fed into the classifiers. For instance, every MRI picture in the Figshare collection is of the Matlab-specified .mat type. Consequently, the image's dimensions need to be expanded to be readable. Next, turn each image into a NumPy array to minimize the amount of space needed for our model. The images are resized to a standardized resolution of 512x512 pixels.

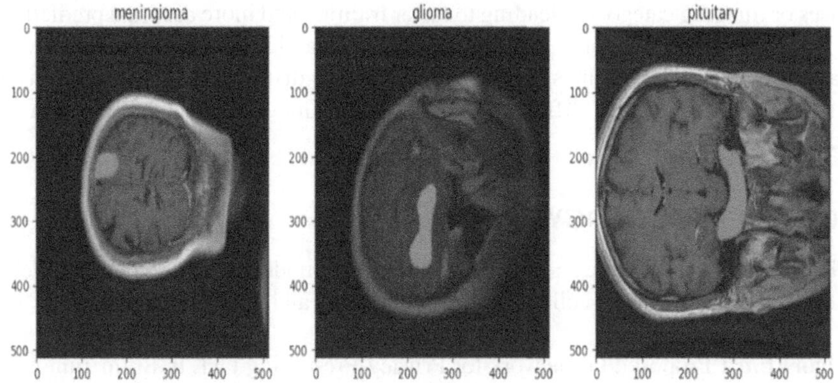

FIGURE 10.4 Brain tumor MRI images.

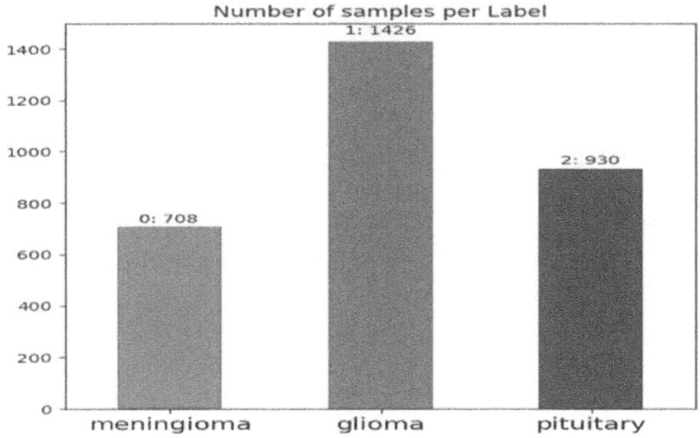

FIGURE 10.5 Label distribution in the Figshare dataset.

Single-channel images are then converted to a three-channel format to match the input requirements of the pre-trained models. The dataset is shuffled and is split into three sections—train, test, and validation. About 80% of the data is used for training, while 20% is used for testing and validation.

10.5.2 FTL Framework

The FTL framework employed in this study is based on the principle of training DL models without exchanging raw data, addressing concerns about data security and privacy in the healthcare industry by keeping data on-site. To achieve our objective of collaborative learning, an architecture setup is required. Figure 10.6 shows the

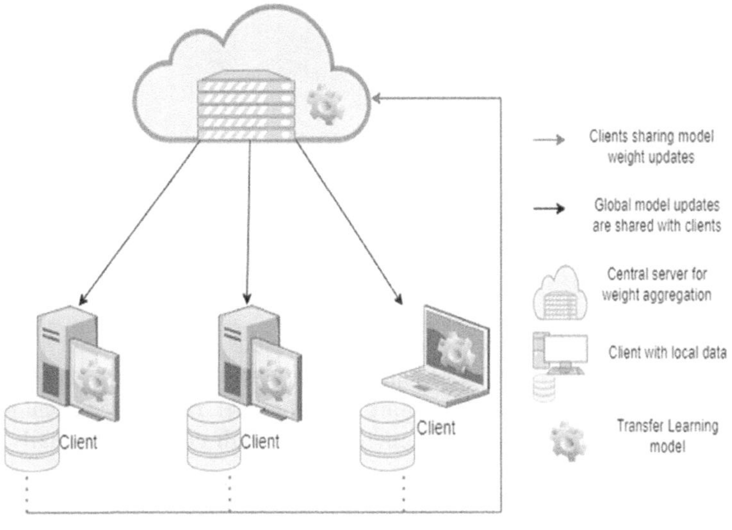

FIGURE 10.6 Proposed FTL architecture.

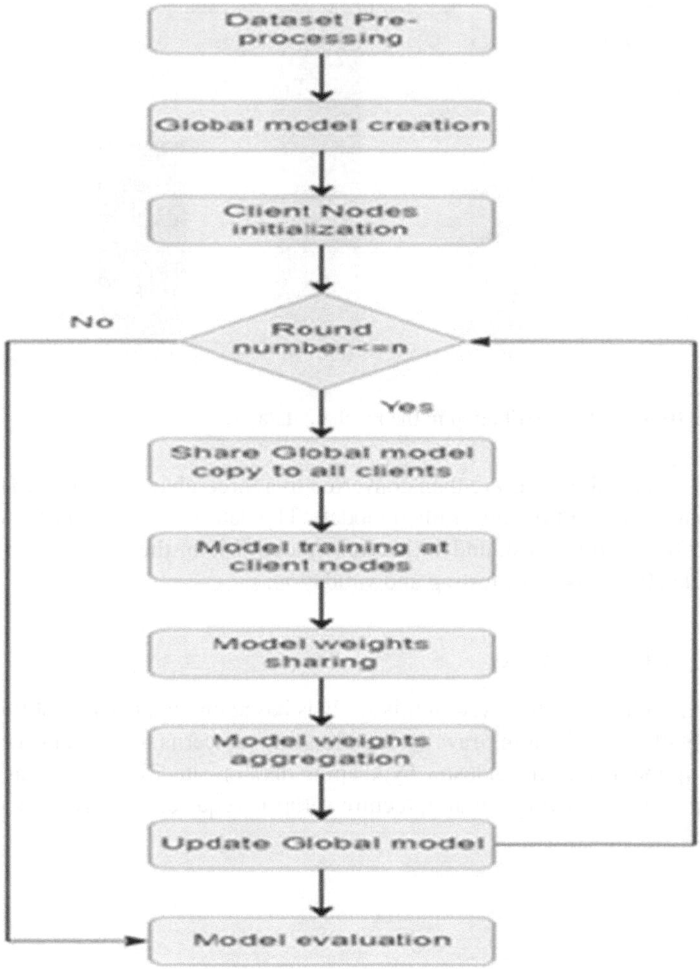

FIGURE 10.7 The proposed model flowchart.

FTL architecture, while Figure 10.7 illustrates the flowchart of the proposed model. Here, a simulated environment is considered, including the following architectural components:

- **Client Nodes:** These nodes represent hospitals, clinics, research institutions, etc., and locally store patient data without sharing it. Each client serves as a data source for training a local model. Here, three client nodes are considered in a simulated environment, with the dataset equally split among them.
- **Global Model:** The global model constitutes the ML/DL algorithm utilized for the classification task. Here, the TL model with fine-tuning is considered

as the global model. In this context, two specific pre-trained models are considered for FTL, namely VGG16 and VGG19 [23]. Both models are known for their simplicity and depth, making them effective for extracting hierarchical features from medical images. A replica of this model is maintained at each client node, referred to as the local model. After each round of weight aggregation, the global model is updated, thereby synchronizing the local models on the client nodes.

- **Central Server**: Also known as the aggregation server, hosts the global model and facilitates the aggregation of model weights from client nodes. It manages the training iterations and updates the global model with aggregated weights [24, 25].

Training procedure: It involves multiple rounds of FL, with each round comprising training and aggregation phases [26, 27]. In each round, client nodes train their local models on their local datasets for a fixed number of epochs. Following training, the client nodes transmit their model weights to the central server, where FedAvg is used to aggregate the weights using the formula:

$$w_{t+1} = \sum_{k=1}^{K} \frac{n_k}{n} . w_k \tag{10.1}$$

where

- w_{t+1} is the updated global model at round $t + 1$
- w_k is the local model trained by client k
- n_k is the number of samples held by client k
- n is the total number of samples across all clients

The updated global model is produced by averaging the client's local model weights, which are weighted according to the percentage of data that each client has. This way, the final global model is obtained from the contribution of each client, proportionate to the amount of its local dataset. This guarantees that the flow of information remains complete throughout the FL process while ensuring data privacy [28].

After each round, the performance of the global model is assessed to track its accuracy. Additionally, local model accuracy is assessed on each client node to evaluate the performance of each node in the FL process. After a predetermined number of rounds, the final aggregated weights are updated as the global model in the central server.

10.6 EXPERIMENTS AND RESULTS

10.6.1 Experimental Setup

The experimental setup involved the utilization of a DGX GPU server equipped with high-performance NVIDIA GPUs to conduct simulations of FTL without relying on any existing FL frameworks. The objective was to develop a custom FL framework using Python version 3.8 and TensorFlow within a Jupyter Notebook environment. The hardware used in our experiments is Intel Core i9 CPUs with 8 cores and 64 (2 x 32GB) DDR4RAM and Nvidia RTX 3060 GPUs with 12 GB GDDR6 graphics. The GPU server provided significant computational power and memory resources necessary for training FTL models in a simulated environment.

10.6.2 Implementation

An FTL framework in a simulated environment with 3 client nodes is considered for brain tumor multi-class classification. Here, we considered IID data to have a similar data distribution at all the client nodes. The dataset is randomly shuffled and is equally distributed among the 3 clients after splitting the dataset using the eighty-twenty rule for training and testing.

A global FTL model is constructed by freezing the fully connected layers of pre-trained models, specifically VGG16 and VGG19 [23]. Customized fine-tuning is then applied to adapt these pre-trained models to the Brain MRI dataset, incorporating different fully connected layers with 1024, 1024, 512, and 3 neurons each. The last layer utilizes a softmax activation function for multi-classification, while ReLU activation functions are employed for preceding layers. Additionally, a dropout of 0.2 is implemented to enhance model generalizability and mitigate overfitting by randomly dropping 20% of the neurons during training. The training process is iterated for five rounds.

10.6.3 Results and Analysis

Table 10.2 presents the performance of the FTL framework using VGG16 and VGG19 models. Along with accuracy, other important evaluation metrics such as F1-score, precision, and recall are reported to give a more complete picture of model effectiveness [29]. The confusion matrices, shown in Figure 10.8, illustrate how well the models are distinguishing between classes. A direct comparison between VGG16

TABLE 10.2
Comparison of VGG16 and VGG19 Evaluation Metrics

FTL Model	Accuracy (%)	Class Label	Precision (%)	Recall (%)	F1-Score (%)
VGG16	93.11	Meningioma	84	85	85
		Glioma	96	94	95
		Pituitary	94	99	96
VGG19	91.8	Meningioma	93	69	79
		Glioma	90	98	94
		Pituitary	95	97	96

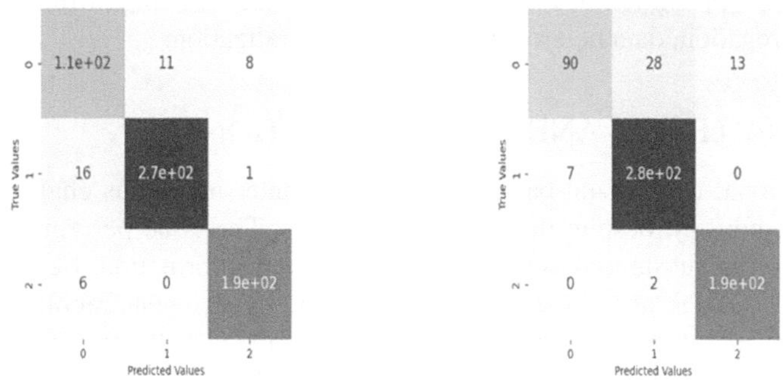

FIGURE 10.8 Confusion matrices for VGG16 and VGG19.

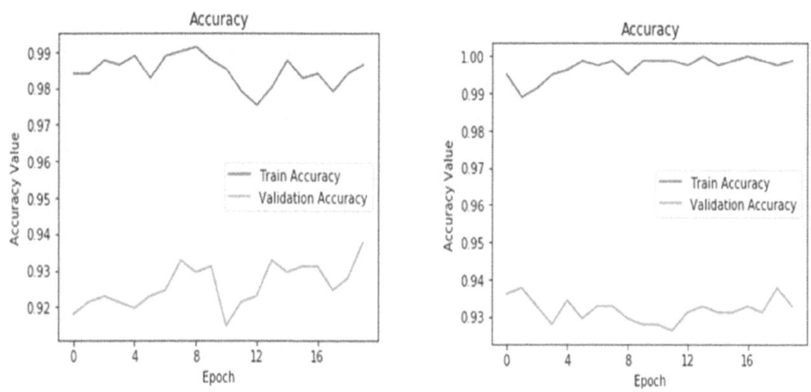

FIGURE 10.9 Performance comparison of VGG16 and VGG19.

and VGG19 is provided in Figure 10.9, which also indicates that neither model is overfitting, as training and validation results remain consistent. VGG16 achieved an accuracy of 93.11%, slightly outperforming VGG19, which reached 91.8%. This suggests that VGG16 is more appropriate compared to VGG19. Overall, the results demonstrate that the FTL framework, by combining FL and TL, ensured better model performance without compromising data privacy.

10.7 CONCLUSION

To conclude, the use of VGG16 and VGG19 architectures in FTL for brain tumor classification produced good results with corresponding accuracy rates of 93.11% and 91.8% respectively. This shows how well the FTL technique uses pre-trained models for medical image analysis. The results highlight FTL as a reliable and effective framework for decentralized ML tasks, especially in fields such as healthcare. Further studies are required to investigate the scalability, generalizability, and practical application of FTL frameworks in clinical settings while considering secure aggregation, data heterogeneity, and decentralization.

10.8 CHALLENGES AND FUTURE DIRECTIONS

A Conventional centralized-based methods encounter numerous challenges in the healthcare industry, despite the advantages of TL. The issue pertains to data privacy due to the substantial volume of health-related information. Legislation such as Europe's GDPR and America's HIPAA imposes stringent regulations on the handling of this data. However, in fact, the centralized collection of patient data for training purposes may pose risks of privacy infringements and potential data breaches. A significant difficulty is the heterogeneity of data; medical pictures can differ substantially between institutions due to variations in acquisition techniques, equipment, and patient demographics. No one desires APC models that have been solely trained on regular or homogeneous data to underperform across diverse clinical scenarios. Moreover, healthcare datasets are often characterized by class imbalance and distribution shifts, meaning that some diseases are represented by fewer samples or derived from non-uniformly sampled populations. The consequence of this may be models that are biased and have poor generalization to new scenarios or less common cases. By solving these problems, FL will be more useful and powerful, especially in healthcare, where both protecting patient privacy and improving care are very important.

REFERENCES

1. Khan, M.S.I., Rahman, A., Debnath, T., Karim, M.R., Nasir, M.K., Band, S.S., Mosavi, A. and Dehzangi, I. 2022. Accurate brain tumor detection using deep convolutional neural network. *Computational and Structural Biotechnology Journal*, 20, pp. 4733–4745.
2. Anjum, S., Hussain, L., Ali, M., Alkinani, M.H., Aziz, W., Gheller, S., Abbasi, A.A., Marchal, A.R., Suresh, H. and Duong, T.Q. 2022. Detecting brain tumors using deep learning convolutional neural network with transfer learning approach. *International Journal of Imaging Systems and Technology*, 32(1), pp. 307–323.
3. McMahan, B., Moore, E., Ramage, D., Hampson, S. and y Arcas, B.A. 2017, April. Communication-efficient learning of deep networks from decentralized data. In *Artificial Intelligence and Statistics* (pp. 1273–1282). PMLR.
4. Kone˘cny´, J., McMahan, H.B., Yu, F.X., Richt´arik, P., Suresh, A.T. and Bacon, D. 2016. Federated learning: Strategies for improving communication efficiency. *arXiv preprint arXiv*:1610.05492.
5. Islam, T.U., Ghasemi, R. and Mohammed, N. 2022, January. Privacypreserving federated learning model for healthcare data. In *2022 IEEE 12th Annual Computing and Communication Workshop and Conference (CCWC)* (pp. 0281–0287). IEEE.
6. Khan, L.U., Pandey, S.R., Tran, N.H., Saad, W., Han, Z., Nguyen, M.N. and Hong, C.S. 2020. Federated learning for edge networks: Resource optimization and incentive mechanism. *IEEE Communications Magazine*, 58(10), pp. 88–93.
7. Javeed, D., Saeed, M.S., Kumar, P., Jolfaei, A., Islam, S. and Islam, A.N. 2023. Federated learning-based personalized recommendation systems: An overview on security and privacy challenges. *IEEE Transactions on Consumer Electronics*, 70(1), pp.2618-2627.

8. Nguyen, D.C., Ding, M., Pathirana, P.N., Seneviratne, A., Li, J., Niyato, D. and Poor, H.V. 2021. Federated learning for industrial internet of things in future industries. *IEEE Wireless Communications*, 28(6), pp. 192–199.
9. Islam, M., Reza, M.T., Kaosar, M. and Parvez, M.Z. 2023. Effectiveness of federated learning and CNN ensemble architectures for identifying brain tumors using MRI images. *Neural Processing Letters*, 55(4), pp. 3779–3809.
10. Tedeschini, B.C., Savazzi, S., Stoklasa, R., Barbieri, L., Stathopoulos, I., Nicoli, M. and Serio, L. 2022. Decentralized federated learning for healthcare networks: A case study on tumor segmentation. *IEEE Access*, 10, pp. 86938708.
11. Elmas, G., Dar, S.U., Korkmaz, Y., Ceyani, E., Susam, B., Ozbey, M., Avestimehr, S. and C̦ukur, T. 2022. Federated learning of generative image priors for MRI reconstruction. *IEEE Transactions on Medical Imaging*, 42(7), pp.1996-2009.
12. Zhang, M., Qu, L., Singh, P., Kalpathy-Cramer, J. and Rubin, D.L. 2022. Splitavg: A heterogeneity-aware federated deep learning method for medical privacy-preserving FTL for brain tumor classification in healthcare imaging. *IEEE Journal of Biomedical and Health Informatics*, 26(9), pp. 46354644.
13. Nazir, S. and Kaleem, M. 2023. Federated learning for medical image analysis with deep neural networks. *Diagnostics*, 13(9), p. 1532.
14. Canetti, R., Lindell, Y., Ostrovsky, R. and Sahai, A. 2002, May. Universally composable two-party and multi-party secure computation. In *Proceedings of the Thiry-Fourth Annual ACM Symposium on Theory of Computing* (pp. 494–503).
15. Dwork, C. and Roth, A. 2014. The algorithmic foundations of differential privacy. *Foundations and Trends® in Theoretical Computer Science*, 9(3–4), pp. 211–407.
16. Truex, S., Baracaldo, N., Anwar, A., Steinke, T., Ludwig, H., Zhang, R. and Zhou, Y. 2019, November. A hybrid approach to privacy-preserving federated learning. In *Proceedings of the 12th ACM Workshop on Artificial Intelligence and Security* (pp. 1–11).
17. Viet, K.L.D., Le Ha, K., Quoc, T.N. and Hoang, V.T. 2023, May. MRI brain tumor classification based on federated deep learning. In *2023 Zooming Innovation in Consumer Technologies Conference (ZINC)* (pp. 131–135). IEEE.
18. Rana, O., Spyridopoulos, T., Hudson, N., Baughman, M., Chard, K., Foster, I. and Khan, A. 2022, December. Hierarchical and decentralised federated learning. In *2022 Cloud Continuum* (pp. 1–9). IEEE.
19. Yang, Q., Liu, Y., Chen, T. and Tong, Y. 2019. Federated machine learning: Concept and applications. *ACM Transactions on Intelligent Systems and Technology (TIST)*, 10(2), pp. 1–19.
20. Alanazi, M.F., Ali, M.U., Hussain, S.J., Zafar, A., Mohatram, M., Irfan, M., AlRuwaili, R., Alruwaili, M., Ali, N.H. and Albarrak, A.M. 2022. Brain tumor/mass classification framework using magnetic-resonance-imaging-based isolated and developed transfer deep-learning model. *Sensors*, 22(1), p. 372.
21. Shin, H.C., Roth, H.R., Gao, M., Lu, L., Xu, Z., Nogues, I., Yao, J., Mollura, D. and Summers, R.M. 2016. Deep convolutional neural networks for computer-aided detection: CNN architectures, dataset characteristics and transfer learning. *IEEE Transactions on Medical Imaging*, 35(5), pp. 12851298.
22. Figshare dataset. URL: https://figshare.com/articles/braintumordataset/1512427.
23. Simonyan, K. and Zisserman, A. 2014. Very deep convolutional networks for large-scale image recognition. *arXiv preprint arXiv*:1409.1556.
24. Truhn, D., Arasteh, S.T., Saldanha, O.L., Mu¨ller-Franzes, G., Khader, F., Quirke, P., West, N.P., Gray, R., Hutchins, G.G., James, J.A. and Loughrey, M.B. 2024. Encrypted federated learning for secure decentralized collaboration in cancer image analysis. *Medical Image Analysis*, 92, p. 103059.

25. Lytvyn, O. and Nguyen, G. 2023. Secure multi-party computation for magnetic resonance imaging classification. *Procedia Computer Science*, 220, pp. 24–31. 18 Privacy-Preserving FTL for Brain Tumor Classification in Healthcare
26. Saldanha, O.L., Quirke, P., West, N.P., James, J.A., Loughrey, M.B., Grabsch, H.I., Salto-Tellez, M., Alwers, E., Cifci, D., Ghaffari Laleh, N. and Seibel, T. 2022. Swarm learning for decentralized artificial intelligence in cancer histopathology. *Nature Medicine*, 28(6), pp. 1232–1239.
27. Warnat-Herresthal, S., Schultze, H., Shastry, K.L., Manamohan, S., Mukherjee, S., Garg, V., Sarveswara, R., H¨andler, K., Pickkers, P., Aziz, N.A. and Ktena, S. 2021. Swarm learning for decentralized and confidential clinical machine learning. *Nature*, 594(7862), pp. 265–270.
28. Korkmaz, C., Kocas, H.E., Uysal, A., Masry, A., Ozkasap, O. and Akgun, B. 2020, November. Chain fl: Decentralized federated machine learning via blockchain. In *2020 Second International Conference on Blockchain Computing and Applications (BCCA)* (pp. 140–146). IEEE.
29. Pradhan, A.K., Dhulipalla, S.H., Nishat, S.T. and Ghanta, S. 2025. Smart and effective healthcare for diabetic patients using ML techniques. In *Real World Applications and Implementations of IoT* (pp. 213–229). Singapore: Springer Nature Singapore.

11 Futuristic Health Care Management System Using AI

Bhavana Sachdeva, Sonam Sirohi, Arun Kumar Singh, Sandeep Saxena, and Debabrata Samanta

11.1 INTRODUCTION

Artificial intelligence (AI) is an extensive term that incorporates a variety of technologies that replicate the fundamental processes of human intelligence [1, 2]. These systems consider sensory perception, engagement, deep learning, cognition, and adaptability. More and more evidence suggests that clinical research and treatment are using artificial intelligence of some kind. This study has shown the significance of AI-based health solutions and the benefits they provide. It is currently a trend among governments and innovative firms to concentrate resources across the board to the application of AI for healthcare applications. This trend is expected to continue in the near future.

AI has emerged as a disruptive force across a variety of different businesses, and the healthcare industry is at the forefront of adopting AI into its operations. AI has a huge and bright future in the healthcare business. It has the potential to change patient care, diagnosis, treatment, and the administration of healthcare as a whole [3]. In the process of bringing about the future of AI in the healthcare [4] business, this study explores the potential, obstacles, and influence that specialized education programs such as the postgraduate degree in AI have during the process of bringing about the future of AI.

In the healthcare industry, AI is being used for a variety of different elements, and it is also being coupled with other technologies that are connected to it. Deep learning, machine learning, and natural language processing are some of the technologies that fall under this category.

11.1.1 Types of AI of Relevance to Healthcare

Instead of being a single technology, AI is actually a collection of technologies that work together to create a single system [5–8]. Although most of these technologies

DOI: 10.1201/9781003567585-11

have an immediate significance in the healthcare industry, the specific processes and tasks that they assist are quite diverse. Despite this, the majority of these technologies have an immediate significance. The following is an outline and explanation of several AI technologies that are of considerable value to the healthcare business.

11.1.2 Machine Learning—Neural Networks and Deep Learning

In the field of statistics, machine learning is a technique that combines fitting models to data and 'learning' by training models with data. This technique is also known as "machine learning." In addition to that, it is also known as "machine learning." In 2018, Deloitte conducted a survey that was given to 1,100 managers in the United States who were already experimenting with AI in their respective companies. The survey was presented to those managers who were already testing the waters. A total of sixty-three percent of the firms that were investigated were using machine learning in their operations, as indicated by the findings of the survey [9–11]. The term "machine learning" refers to one of the most prevalent forms of AI, although there are several categories of AI. The foundation for a wide range of alternative methods to AI is this method, which may be modified in a wide variety of ways. These alterations can be made to this method.

Neural networks are an example of a technique that falls under the category of machine learning that is more complex. For a considerable number of decades, this technique has been firmly entrenched in the field of study pertaining to healthcare. The 1960s marked the beginning of its existence, and it has been around for a considerable amount of time. Applications that involve categorization have been carried out with the help of neural networks. One example of this is the determination of whether or not a patient would develop a specific illness. Problems are evaluated via the lens of these three concepts: the inputs, the outputs, and the weights of variables [12–15] or "features" that are responsible for associating inputs with outputs. Inputs, outputs, and the weights of variables are the three concepts that are considered. It has been established that there is a connection between it and the manner in which neurons process signals; nevertheless, the comparison to the function of the brain is not very strong. Furthermore, deep learning is increasingly being used for speech recognition, and as such, it refers to a form of natural language processing (NLP) [16], which will be discussed further down in this chapter on the next page. A human observer, in contrast to more conventional methods of statistical analysis, does not typically attribute a great deal of significance to each individual feature that is integrated into a deep learning model. This is because deep learning models are designed to learn from their own data. As a consequence of this, the interpretation of the explanation of the model's outputs might be extremely challenging, and it may even be thought impossible to accomplish.

11.1.3 Natural Language Processing

Researchers in the area of AI have been working to comprehend the language that we all speak since the 1950s. Research in the area of natural language processing

(NLP) offers a wealth of potential new applications. Speech recognition, text analysis, translation, and other similar technologies form the backbone of natural language processing (NLP) [17]. Semantic natural language processing and statistical natural language processing are the two main tools for this job. Many of the recent improvements in recognition performance have their roots in machine learning, particularly deep learning neural networks, which constitute the foundation of statistical natural language processing (NLP). Machine learning is the cornerstone of NLP. It is possible that natural language processing will considerably enhance recognition accuracy.

11.1.4 Rule-Based Expert Systems

During the 1980s, the most widely used type of AI technology comprised expert systems that were constructed using collections of "if-then" rules. Additionally, throughout that time period and in the years that followed, these systems were found to be widely implemented in enterprise settings. Over the course of the last couple of decades, they have been deployed extensively in the field of healthcare for the purpose [18] of providing "clinical decision support," and they continue to be utilized extensively in the current day on a regular basis. In today's market, the vast majority of businesses that provide electronic health record (EHR) systems include a set of recommendations with their products. The participation of individual experts and knowledge engineers is required for the aim of developing a set of rules in a particular knowledge domain. This is because expert systems are designed to accomplish this goal.

11.1.5 Physical Robots

The fact that more than 200,000 industrial robots are installed every year all over the world has contributed to the fact that physical robots are already well-known at this point in time. In settings such as factories and warehouses, they are accountable for carrying out planned actions such as lifting, transferring, welding, or assembling things. Additionally, they are responsible for transporting supplies in a variety of hospitals. Robots have recently become more cooperative with people and are easier to educate by being guided through a desired activity [19]. This has taken place in recent years. The field of robotics has been opened up to new possibilities as a result of this. However, in addition to this, they are gaining intelligence as a result of the incorporation of more capabilities of AI into their "brains," which are actually their operating systems. It is quite conceivable that the equivalent improvements in intelligence that we have experienced in other areas of AI will eventually be applied to physical robots. This is something that will happen over the course of time.

11.1.6 Robotic Process Automation

By utilizing this technology, organized digital chores for administrative purposes, such as those involving information systems, are carried out in the same manner as if they were carried out by a human user who adhered to a script or set of rules [20]. This is the case because the technology is able to perform these activities in the same manner. In contrast to other forms of AI, they are not only inexpensive but also easy to create and transparent in their behavior. In addition, they are surprisingly easy to understand. As a matter of fact, robotic process automation (RPA) does not involve any robots at all; rather, it is made up of computer programs that are executed on servers. When it comes to becoming a semi-intelligent user of information systems, it is dependent on a mix of workflow, business rules, and interaction with the "display layer" of the systems. This interaction is made possible by the combinations of these three factors. When it comes to the healthcare industry, it is usual practice to use them for positions that involve repeated tasks, such as prior authorization, updating patient information, or billing.

11.2 AI IN HEALTHCARE

In the event that AI were to be implemented, there is a possibility that it would significantly improve the effectiveness of healthcare systems [21] as well as the outcomes for patients. When it comes to healthcare, some of the uses of AI include the following .

11.2.1 Diagnosis and Disease Prediction

These algorithms are powered by AI and are capable of analyzing large datasets, which enables them to assist in the prediction of diseases and the development of more accurate diagnoses. Machine learning algorithms can be employed to predict the potential disease based on the data collected from the patients [22]. In order to better tailor treatments to individual patients, an increasing number of doctors are turning to AI. Therefore, we emphasize that AI has the capacity to learn from data from previous encounters in order to provide patients with more effective treatment options and to contribute to the evaluation of treatment strategies [17]. This enables it to deliver better care for its patients. Among the many well-known uses of ML in digital imaging, medical imaging has made heavy use of it for computer-aided detection (CAD) and picture interpretation. A fresh strategy is the direct implementation of machine learning. Medical imaging data generated by X-ray, CT, MRI, and ultrasound scanners may be enhanced by AI to detect illness indicators more rapidly and with greater precision [23]. Doing this may make things go more smoothly. One of the benefits for patients is that they can get more treatments and get a faster and more accurate analysis of their disease.

11.2.2 Treatment Personalization

These AI-powered algorithms can analyze massive datasets, enabling them to help with illness prediction and diagnosis. Machine learning algorithms can forecast illness based on patient data. Physicians are increasingly adopting AI to create patient treatment strategies. We highlight AI because it can learn from prior interactions to provide patients improved treatment modalities and assess treatment tactics [17]. It can treat people better. ML has been widely utilized in medical imaging for computer-aided detection/cad and picture interpretation. Directly using machine learning is new. AI can examine CT, MRI, X-ray, and ultrasound scans to detect illness quicker and more precisely. This may ease the procedure. Patients gain from additional therapy options and quicker, more accurate diagnosis.

11.2.3 Drug Discovery

AI helps to speed up the process of drug development by performing tasks such as the analysis of molecular structures, the prediction of possible drug candidates, and the optimization of clinical trial designs [24]. The large chemical space, which is made up of more than 1060 different compounds, provides an environment that is favorable to the development of a significant number of therapeutic molecules. The process of developing new drugs, on the other hand, is complicated by the lack of development of new technologies.

11.2.4 Virtual Health Assistants

Chatbots and virtual health assistants, which are able to deliver prompt responses to inquiries posed by patients, arrange appointments, and offer medical advice, can be utilized to enhance patient involvement. This can be accomplished through the utilization of these technologies. Virtual healthcare assistants are comparable to intelligent assistants in that they make the lives of both patients and physicians easier. These assistants are powered by AI. Patients have the ability to plan appointments, it is possible to provide answers to their core health issues, and it is possible to remind patients to take their prescription [25–28]. Due to the fact that they are always prepared and operating around the clock, individuals are able to receive support whenever they require it. Following the analysis and processing of the inputs that are provided by users, solutions are then provided to the user in the form of either text or speech, depending on the questions that the user might have asked.

Medical personnel are able to dedicate a greater portion of their focus to the treatment of patients as a result of this possibility. It is intriguing for students to witness how technological improvements are revolutionizing healthcare by making it more efficient and effective while still preserving its characteristic of being individualized and caring.

11.2.5 Healthcare Operations

AI facilitates the integration of administrative responsibilities, including the administration of medical records, the establishment of appointments, and the invoicing process. In turn, this reduces administrative weariness and costs. The healthcare industry is now the most prominent user of AI [29, 30] due to the need for both speed and accuracy in the medical field. For instance, consider the following AI applications in this field right now: Medical imaging data, such as X-rays, MRI, and CT scan findings, may be evaluated using algorithms driven by AI. In order to help doctors make faster and more accurate diagnoses, several algorithms have been developed.

11.3 USE OF AI IN HEALTHCARE

Acquire an improved awareness of the various applications of AI in healthcare-related scenarios. AI can improve healthcare in many ways, such as by automating administrative tasks, supporting clinical decision-making, and making the healthcare system more efficient and effective across a wide range of areas. These benefits should be looked into in more detail.

11.3.1 Telemedicine and Telehealth

AI has transformed the area of telemedicine by making it feasible to conduct appointments remotely using chatbots, virtual assistants, and other tools [31]. Medical practitioners may provide basic medical advice and early examinations without requiring patients to leave their homes. Furthermore, AI is a powerful tool for remotely monitoring patients' health by analyzing data received from wearables and sensors. Remote monitoring stands out as a critical component in enabling this. Medical personnel are able to personalize treatment plans to the specific patient and act in a proactive manner as a result. AI also aids decision-making and diagnosis by analyzing electronic health records (EHRs) and medical imaging for abnormalities and trends.

11.3.2 Robotics in Surgical Procedures

There are several advantages of using AI-robots to perform the surgery:

- Precision Robotic systems are more precise and accurate than the human hand, which reduces human error and allows for fine movements in procedures.
- AI is leading to less invasive surgery, meaning smaller incisions, less pain and a shorter recovery as well as better cosmetic results.
- Real-time feedback and safety guidelines enhance patient safety during surgeries, which can help prevent errors and risks.
- Less-invasive surgeries result in faster and more comfortable recoveries, as patients can return more quickly to their normal routines.
- AI assists in developing personalized surgical plans tailored to the anatomy and condition of each patient.
- Remote surgery might be helped along by advances in communications technology.

Robotic AI is revamping surgery, delivering safety and superior outcomes to patients.

11.3.3 AI for Mental Health Diagnosis

The field of mental diseases is greatly impacted by AI in terms of both diagnosis and therapy. AI systems have the potential to improve our understanding of mental health diseases by identifying patterns and offering insights from massive amounts of data. The mental health care system is poised for a sea change if this technology is widely used for the diagnosis, treatment, and prevention of mental illness. AI's capacity to accurately and rapidly analyze massive information is one of its primary benefits to mental health evaluation [32]. The reasoning for this is because these professionals have access to more data when deciding on diagnostic and treatment methods. Another factor is that AI can also identify those who are at risk for developing mental health issues. Prevention and early intervention strategies become feasible in more realistic cases. AI can also identify depressed individuals.

11.3.4 Predictive Analytics in Preventive Healthcare

The field of preventative healthcare stands to benefit greatly from the advancements in predictive analytics, which are connected to AI. With the use of AI, we can better analyze patterns and risk factors to identify those most at risk of developing particular illnesses. By examining trends, we can do this. Because of this, patients have been able to get treatment sooner, which has reduced strain on healthcare systems and improved their overall health. Making adjustments to one's way of life and undergoing targeted medical treatment are two of the many options available for reducing the likelihood of getting major illnesses later in life.

11.3.5 Future of AI in Healthcare

The future driven by AI could potentially be observed through an anticipation into the healthcare technology crystal ball. In addition to staying current on groundbreaking research and trends, Children's Hospital hopes that you will recognize how AI has the ability to revolutionize healthcare.

11.3.6 AI Ethics and Regulations in Healthcare

In order to use AI, the healthcare business needs rules and guidelines based on ethical principles. Primarily, ethical considerations should center on minimizing prejudice, obtaining patients' agreement, and preserving their privacy. It is necessary to construct regulatory frameworks in the absence of data governance standards, including certification procedures, accountability, and openness [18]. Adhering to these principles will allow us to guarantee that AI can contribute to a future in healthcare that is both responsible and ethical.

11.3.7 Patient Data Privacy and Security

The effective integration of AI into healthcare relies on the secure storage of patient data. Health care providers, consumers, and regulators may all benefit from a more trustworthy relationship thanks to well-executed privacy and security regulations.

11.3.7.1 Privacy Measures

- The process of de-identifying data involves removing direct identifiers such as names and addresses prior to the training or analysis of AI.
- The protection of patient data both while it is at rest and while it is in transit through the use of encryption.
- Through the use of passwords and biometrics, access controls are implemented to restrict access to only authorized persons.

11.3.7.2 Security Measures

- Protecting against external threats such as hackers and viruses is the primary function of firewalls and intrusion detection systems.
- Transmission of patient information across secure networks in order to prevent eavesdropping is an example of secure networks.
- Conducting security audits on a regular basis, with the goal of locating and fixing vulnerabilities in security systems.

For the purpose of ensuring the successful and dependable absorption of AI in the healthcare industry, it is absolutely necessary to put in place strong privacy and security standards.

11.4.1 Applications of AI in Healthcare

Effective disease diagnosis is still regarded a difficulty on a worldwide scale, despite the numerous advancements that have been made in the field of medicine. Because of the intricacy of the many different disease pathways and the symptoms that lie behind them, the development of early diagnostic tools is a continuing challenge. Several facets of medical care, including diagnosis, are ripe for disruption by AI. The field of machine learning (ML) [17–20, 22] is a subfield of AI that makes use of data as an input resource. The accuracy of ML is heavily dependent on the quantity as well as the quality of the data that is input, which can help overcome some of the difficulties and complexities associated with diagnosis. In a nutshell, machine learning has the potential to facilitate decision-making, manage workflow, and automate processes in a manner that is both timely and cost-effective. The addition of layers through the use of Convolutional Neural Networks (CNN) and data mining techniques that assist in the identification of data patterns was another feature of deep learning. The identification of significant disease detection patterns within large datasets is a particularly relevant use of these strategies. These techniques have a wide range of applications in healthcare systems, including disease classification, disease prediction, and disease diagnosis.

As opposed to conventional diagnostic procedures, the use of AI technologies can result in increased precision, decreased expenses, and time savings. Various applications of AI in the field of medicine, including the examination of medical images, X-rays, CT scans, and MRIs, are still being investigated by researchers. AI can also assist in the identification of abnormalities, the detection of fractures, tumors, or other disorders, and the provision of quantitative measures for the purpose of offering a medical diagnosis that is both more accurate and more expedient.

Observable traits that are affected by both genetic and environmental influences can be predicted.

11.4.2 AI Assistance in Population Health Management

Predictive analytics are also being used more and more to focus health tasks in public health management. Data analytics include predictive analytics, which makes use of modeling, data mining, AI, and ML. It does this by analyzing both historical and real-time data in order to make predictions about the future. Machine learning algorithms and other technologies are used to analyze data and create prediction models in order to enhance patient outcomes while decreasing expenses [20, 22]. One area where predictive analytics might be very useful is in determining who is most likely to have a chronic condition, such as diabetes, heart disease, or endocrine disorders. A higher risk patient for a certain disease may be identified using predictive models that are based on data such as patients' medical histories, demographics, and lifestyle variables. On the other hand, certain models could be able to zero in on specific therapies that might cure or prevent these diseases. Hospital readmissions may also be anticipated with the use of predictive analytics. By using patient demographics, medical history, and social health factor data, machine learning may even predict which patients are more likely to have a greater risk of hospital readmissions and then focus therapies on these patients. One of the many reasons why new businesses like "Reveal" are springing up is because it has the potential to improve health outcomes while simultaneously reducing healthcare costs.

11.5 CONCLUSION

The utilization of AI for decision-making is progressively gaining traction in the healthcare sector. These technologies enable health care practitioners to enhance patient outcomes by facilitating data-driven, informed decisions. We can develop superior diagnostic while therapeutic tools based on their astute recommendations. The passive capability of real-time assistance, integrated with various data sources, minimizes mistakes and facilitates time-sensitive situations. This also had advantageous outcomes, including enhanced patient safety, less diagnostic mistakes, and improved treatment planning. The use of AI-based decision support systems is revolutionizing the way healthcare is provided. Several hospitals' radiology, oncology, and cardiology departments have implemented the system. It is envisaged that as AI continues to grow, these technologies will be utilized in increasingly sophisticated ways.

REFERENCES

1. A. D. Godley, trans. *Herodotus, Histories.* Cambridge: Harvard University Press, 1920.
2. Endalamaw, A., Gilks, C.F., Ambaw, F. and Assefa, Y., 2022. Universality of universal health coverage: A scoping review. PloS one, 17(8), p.e0269507.
3. Villarreal, Edgar R. Dulce, Jose García-Alonso, Enrique Moguel, and Julio Ariel Hurtado Alegría. "Blockchain for healthcare management systems: A survey on interoperability and security." *IEEE Access* 11 (2023): 5629–5652.
4. Sheeba, A., and A. Seilan. "Health care system in India: An overview." *International Issues on Health Economics and Management* (2010): 215–218.
5. Aggelidis, Vassilios P., and Prodromos D. Chatzoglou. "Using a modified technology acceptance model in hospitals." *International Journal of Medical Informatics* 78, no. 2 (2009): 115–126.
6. Da Costa, Cristiano André, Cristian F. Pasluosta, Björn Eskofier, Denise Bandeira Da Silva, and Rodrigo da Rosa Righi. "Internet of Health Things: Toward intelligent vital signs monitoring in hospital wards." *Artificial Intelligence in Medicine* 89 (2018): 61–69.
7. Fan, Yimeng, Pengyun Xu, Hao Jin, Jianlong Ma, and Liang Qin. "Vital sign measurement in telemedicine rehabilitation based on intelligent wearable medical devices." *Ieee Access* 7 (2019): 54819–54823.
8. Khatoon, Asma. "A blockchain-based smart contract system for healthcare management." *Electronics* 9, no. 1 (2020): 94.
9. Sidhu, R. K. "Health care in India: critical analysis." *International Journal of Medical and Health Sciences* 4, no. 2 (2015): 264–268.
10. Kala, Chandra Prakash. "Traditional health care systems and herbal medicines." *European Journal of Environment and Public Health* 1, no. 1 (2017): 03.
11. Kala, Chandra Prakash, Pitamber Prasad Dhyani, and Bikram Singh Sajwan. "Developing the medicinal plants sector in northern India: challenges and opportunities." *Journal of Ethnobiology and Ethnomedicine* 2 (2006): 1–15.
12. Singh, M. and Jyani, G., "Indian health system & healthcare financing." *A Handbook of Health System Costing*, pp.17–26.
13. Chauhan, L. S. "Public health in India: issues and challenges." *Indian Journal of Public Health* 55, no. 2 (2011): 88–91.
14. Preetha, G. S. *Private Healthcare Sector–A Vibrant Force in Indian Healthcare Industry.* IIHMR: New Delhi, 2019.
15. Pawar, P., Parolia, N., Shinde, S. *et al.* "eHealthChain—a blockchain-based personal health information management system." *Annals of Telecommunications* **77** (2022): 33–45 . https://doi.org/10.1007/s12243-021-00868-6.
16. Minopoulos, Georgios M., Vasileios A. Memos, Christos L. Stergiou, Konstantinos D. Stergiou, Andreas P. Plageras, Maria P. Koidou, and Konstantinos E. Psannis. "Exploitation of emerging technologies and advanced networks for a smart healthcare system." *Applied Sciences* 12, no. 12 (2022): 5859.
17. Nahavandi, Darius, Roohallah Alizadehsani, Abbas Khosravi, and U. Rajendra Acharya. "Application of artificial intelligence in wearable devices: Opportunities and challenges." *Computer Methods and Programs in Biomedicine* 213 (2022): 106541.
18. Manickam, Pandiaraj, Siva Ananth Mariappan, Sindhu Monica Murugesan, Shekhar Hansda, Ajeet Kaushik, Ravikumar Shinde, and S. P. Thipperudraswamy. "Artificial intelligence (AI) and internet of medical things (IoMT) assisted biomedical systems for intelligent healthcare." *Biosensors* 12, no. 8 (2022): 562.

19. Kamruzzaman, M. M., Ibrahim Alrashdi, and Ali Alqazzaz. "New opportunities, challenges, and applications of edge-AI for connected healthcare in internet of medical things for smart cities." *Journal of Healthcare Engineering* 2022, no. 1 (2022): 2950699.
20. Jabbar, Rateb, Noora Fetais, Moez Krichen, and Kamel Barkaoui. "Blockchain technology for healthcare: Enhancing shared electronic health record interoperability and integrity." In *2020 IEEE International Conference on Informatics, IoT, and Enabling Technologies (ICIoT)*, pp. 310–317. IEEE, 2020.
21. Efthymiou-Egleton, Iris-Panagiota, Symeon Sidiropoulos, Dimitrios Kritas, Athanassios Vozikis, Paraskevi Rapti, and Kyriakos Souliotis. "AI transforming Healthcare Management during Covid-19 pandemic." *HAPSc Policy Briefs Series* 1, no. 1 (2020): 130–138.
22. Giri, Arunangshu, Satakshi Chatterjee, Pradip Paul, Smita Chakraborty, and Swatee Biswas. "Impact of smart applications of IoMT (internet of medical things) on healthcare domain in India." *International Journal of Recent Technology and Engineering (IJRTE)* 8, no. 4 (2019): 1–15.
23. Karthick, G. S., and P. B. Pankajavalli. "A review on human healthcare internet of things: a technical perspective." *SN Computer Science* 1, no. 4 (2020): 198.
24. Amin, Syed Umar, and M. Shamim Hossain. "Edge intelligence and Internet of Things in healthcare: A survey." *IEEE Access* 9 (2020): 45–59.
25. Panch, Trishan, Peter Szolovits, and Rifat Atun. "Artificial intelligence, machine learning and health systems." *Journal of Global Health* 8, no. 2 (2018): 020303.
26. Basatneh, Rami, Bijan Najafi, and David G. Armstrong. "Health sensors, smart home devices, and the internet of medical things: an opportunity for dramatic improvement in care for the lower extremity complications of diabetes." *Journal of Diabetes Science and Technology* 12, no. 3 (2018): 577–586.
27. de Kleijn, Maria, Mark Siebert, and Sarah Huggett. "Artificial Intelligence: How knowledge is created, transferred and used." Corfu, Greece, 2019. https://repository.ifla.org/items/190b7cff-39e4-4f19-a788-831de474dfc1
28. Tran, Bach Xuan, Giang Thu Vu, Giang Hai Ha, Quan-Hoang Vuong, Manh-Tung Ho, Thu-Trang Vuong, Viet-Phuong La, et al. "Global evolution of research in artificial intelligence in health and medicine: A bibliometric study." *Journal of Clinical Medicine* 8, no. 3 (2019): 360.
29. Alaskar, Kamal M., Firoj A. Tamboli, Shabana A. Memon, Sampada S. Gulavani, Pralhad K. Mudalkar, Vaibhava V. Desai, Prasanna R. Rasal, Prashant G. Tandale, and Dinanath T. Gaikwad. "Artificial Intelligence (AI) in healthcare management." *Journal of Pharmaceutical Negative Results* 13 (2022): 1011–1020.
30. Gillies, Robert J., Paul E. Kinahan, and Hedvig Hricak. "Radiomics: images are more than pictures, they are data." *Radiology* 278, no. 2 (2016): 563–577.
31. Hatt, Mathieu, Florent Tixier, Dimitris Visvikis, and Catherine Cheze Le Rest. "Radiomics in PET/CT: more than meets the eye?." *Journal of Nuclear Medicine* 58, no. 3 (2017): 365–366.
32. Krittanawong, Chayakrit, Anusith Tunhasiriwet, HongJu Zhang, Zhen Wang, Mehmet Aydar, and Takeshi Kitai. "Deep learning with unsupervised feature in echocardiographic imaging." *Journal of the American College of Cardiology* 69, no. 16 (2017): 2100–2101.

12 Open-Access Datasets for Brain–Computer Interfaces Research

Niha Kamal Basha, Christo Ananth, Akhatov Akmal Rustamovich, Nurmamatov Mekhriddin Qahramonovich, Nazarov Fayzullo Makhmadiyarovich, and Rakhmanov Buribay

12.1 INTRODUCTION

A brain–computer interface (BCI) converts brain activity into functional intent, or the intention to alter, move, control, or manipulate something in your surroundings. In other words, BCIs allow for the mind-alone use of a program or an appliance. There are many ways to quantify a brain impulse. They are non-invasive, semi-invasive, and invasive. Non-invasive BCIs take information from sensors placed on or near the head. Recording devices can be implanted without surgery, and there are no unsafe or painful procedures in non-invasive BCIs. There are four different measurements in non-invasive BCI applications to identify brain activity: (1) MRI, (2) MEG, (3) EEG, and (4) fMRI. Electroencephalogram (EEG) [1, 2] is a procedure that records electrical activity in the brain using small, metal disks on the scalp as electrodes. Electrical impulses are the main way that brain cells communicate with each other, never resting even when you do. Magnetoencephalography (MEG) is a procedure that records the magnetic field generated by the electrical activity of the brain. It is done to trace brain function and ascertain the exact site of the beginning of the epileptic seizure. Magnetic resonance imaging (MRI) is an innocuous treatment that produces very detailed images of our brain. They are created by an energy source, radio waves, and a computer. Functional magnetic resonance imaging (fMRI) [2] evaluates blood oxygenation and blood flow variations brought on by neural activity. It is increasingly being employed in therapeutic settings to map language, higher-order neurocognitive abilities, and sensory processing. An application of MRI/fMRI in BCI is broadly categorized into two:

1) Neurofeedback—which allows a subject to deliberately self-regulate their brain activity by receiving feedback on the amount of activation in a particular region.

 DOI: 10.1201/9781003567585-12

2) Brain decoding—which examines MRI/fMRI data from the subject to ascertain intention. The above-mentioned application is implemented in two ways:
   1) Pattern matching—where the input is processed with respect to task-specific Region of Interest
   2) Machine learning—where the input is processed based on brain state classification.

Here, the data-driven approach plays a key role while taking decisions instead of intuition. The results obtained from the above implementation methods are strategically evaluated based on the data analysis and interpretation. In this chapter, a collection of 15 datasets are found in the OpenNeuro platform. This platform was created by the Stanford Center for Reproducible Neuroscience to encourage and enhance data sharing and analysis of raw MRI data, using BIDS to organize and standardize these data. This dataset [3], comprising diverse neurophysiological data, has the potential to significantly impact BCI research and development. Nonetheless, there is a great difficulty in making full use of these datasets. In this subsection, we present the problem statement, research gap, proposed solutions, domains of application of BCI issues, and their most popular datasets in the research, accompanied by a complete dataset description, collection strategy, and pertinent case study. Open-source BCI datasets [4] are invaluable tools, as they provide a vast amount of neurophysiological information. Still, this absence has led to difficulties in navigating the intricacies of data preprocessing and feature engineering, resulting in inconsistent results and limited experiment reproducibility. The BCI research spans a multitude of domains, including assistive technology, neurorehabilitation, communication, and neuroscientific exploration. The specific datasets that are most relevant and impactful vary depending on the application domain. In this chapter 14 datasets that fall under neurofeedback and brain state decoding application are presented in detail. The proposed chapter has been organized into four categories of datasets. Section 12.2 describes OpenNeuro dataset. Section 12.3 describes BCI competition dataset along with its use cases. Section 12.4 describes motor imagery dataset, Section 12.5 describes brain state decoding, and finally Section 12.6 presents the conclusion.

12.2 OPEN NEURO DATASET AND DESCRIPTION

The participants used to collect the data are white, English-speaking, healthy younger and older adults from the Gainesville, Florida, USA area [1]. None of the participants had either a psychiatric or neurological disorder. The data collection period is from August 2013 to October 2014 for 44 healthy younger adults aged 18–31 years, including 48% of females and 43 older adults aged 63–81 years, with 56% of females. All of the study's older women were postmenopausal, whereas all of its younger women were premenopausal. The cognitive status of participants is initially interviewed over telephone for 30 minutes. Later the eligible participants are called to the University of Florida for a 45-minute cognitive test, and saliva and blood samples are collected. Then the participants undergo a 3-hour test, such as structural scan of T1-weighted

(T1w) with the fMRI scan. Dataset [2] includes 187 MRI data from an age range of 16–81 healthy human adults. A cross-sectional study (AgeRisk) of the growth patterns of risk preference, impulsivity, and self-control included the neuroimaging session. The final of three study sessions included a neuroimaging session that included a structural (T1w) scan, functional scans while performing the Balloon Analogue Risk Task (2 runs), and a delay discounting paradigm. Due to time restrictions, 178 out of 187 individuals finished the delay discounting exercise. This dataset was gathered at the University Hospital Basel in Switzerland between 2016 and 2017. The dataset [3] was collected from 30 adults aged between 18 and 32. The stimulus set contained two identical versions of 1,024 images, which either have with human or without human. The images are classified into 10 groups, such as 2 groups for study and 8 groups for test. The fMRI are collected for the purpose of understanding the effect of responses during recognition memory as well as visual detection. Totally 896 images are used for study with 720 ms and 2,406 images for test. During test time, once the response option screen appeared, participants could react (i) crosshair for 320 to 1,760 ms, (ii) stimulus presentation, and (iii) noise mask for 200 ms, (iv) response option screen for 2,160 ms.

In a cross-sectional study [4] of 88 healthy Colombian men, it was characterized the gut microbiota genera and families related to heart rate variability (HRV) and resting-state functional connectivity (rsFC), their interaction effect over the gut microbiota taxa abundance, and the mediation effect of diet in these relationships. Luminance contrast oscillated at 0.05 Hz with a 12 Hz counter-phase flickering radial checkerboard as part of the visual stimulus run. The breath-holding challenge included three cycles of timed breathing followed by 27 seconds of free breathing, with a total of eight repetitions. This dataset [5] consisted of 50 healthy adult volunteers without a history of neurological or mental disorders. The objective is to deliver a normative dataset for the mapping of cerebral vascular reactivity. To handle the data for this, we created CVRmap, a BIDS-compatible pipeline. Three separate patterns of data make up the raw data.

- Anatomical information: Each patient had a conventional T1w picture taken.
- Functional information: fMRI scans with CO_2 inhalation challenges were performed on everyone.
- Physiological information: During the fMRI acquisition, breath CO_2 was measured using a capnograph.

The Phantom of Bern comprises [6] eight identical-session re-scans of T1-weighted MRI that were performed on two healthy volunteers and obtained with various combinations of sequence settings. The subjects have formally consented to the dissemination of this data, together with the original anonymized DICOM files, and have waived any objections to defacing. The data usage agreement listed below specifies the conditions under which usage is allowed. This dataset can be used to research how acquisition factors affect image contrast because DICOM files with all subject-identifying tags but one are given.

This dataset [7] comprises quantitative FDG-PET, BOLD-fMRI, and DWI data collected during rest on a Siemens Biograph mMR 3T PET-MR scanner from two separate cohorts of 20 right-handed, healthy persons. Nine individuals made up the first cohort, which was collected over the course of two sessions—one with their eyes open and the other with them closed—in the within-subject exploration cohort (TUM.exp). Four ladies and 16 guys with a mean age of 43 years participated in this test. The detailed observations on dataset from OpenNeuro Platform have been prescented in Table 12.1.

12.3 BCI COMPETITION DATASET AND DESCRIPTION

12.3.1 BCI Competition III: Dataset IVa

BCI Competition III: Dataset IVa [8] is a component of a BCI competition series initiated to spur the design and assessment of BCI systems. This dataset, IVa, is associated with motor imagery tasks in which subjects are required to perform imagined actions. The dataset comprises multichannel EEG recordings. Thus, it is useful for investigating the brain's activity in relation to motor imagery.

12.3.1.1 Collection Methods

The EEG data in Dataset IVa was gathered through a BrainVision amplifier [9] with 22 EEG channels. During the recording sessions, participants were asked to carry out two different types of motor imagery tasks: left-hand motor imagery and right-hand motor imagery. Each task consisted of imagining a set of autonomous actions of either the left or the right hand. The EEG signals were recorded with a frequency of 250 Hz, which is a relatively high value and offers a good time resolution of the activity of the brain.

12.3.1.2 Use Cases

BCI Competition III: Dataset IVa has already had many applications [10] in the areas of neuroinformatics, machine learning, and BCI development. Some important applications are:

1. **Development of BCI Systems:** This dataset has been used for the development and testing of motor imagery-based BCI systems.
2. **Feature Extraction and Classification:** The dataset has been used to test different algorithms for feature extraction and classification. Researchers have investigated methods for extracting discriminative features from EEG signals and classifying different motor imagery tasks.

12.3.2 BCI Competition IV: Dataset 2a

This dataset focuses on motor imagery tasks, and it consists of EEG recordings that provide data regarding the brain activity related to motor image visualization. BCI

TABLE 12.1
Details of Dataset from OpenNeuro Platform

Reference	Modality	Task	Open Neuro Accession Number	Sessions	Participants	Participants Ages	Size	Files
1	MRI	Resting state scan	ds004725	1	87	18–80	11.06 GB	185
2	MRI	BART, delay discounting	ds004711	1	187	16–81	21.66 GB	1,664
3	fMRI	Study and test	ds004692	1	30	18–32	326.24 GB	3,302
4	MRI	Rest	ds004648	1	88	N/A	9.61 GB	358
5	MRI	Breath-hold, checker, rest	ds004645	1	15	22–42	42.65 GB	507
6	MRI	Gas	ds004604	1	50	21–67	4.42 GB	306
7	MRI	–	ds004560	2	2	NA	202.13 MB	45
8	MRI, PET	Rest	ds004513	2	20	22–52	40.54 GB	353
9	MRI	Retinotopy, catLoc, FOV	ds004489	3	15	19–28	28.68 GB	1219
10	MRI	Checker	ds004478	1	6	24–33	2.33 GB	55
11	PET, MRI	-	ds004401	1	1	23–42	163.03 MB	7
12	MRI	ImAcq, VisAcq	ds004393	1	32	NA	11.33 GB	2658
13	MRI	Expo, rest	ds004349	1	57	18–28	43.95 GB	1016

Competition IV: Dataset 2a [11] is one of the subsets of the BCI IV Competition series aimed towards pushing forward the growth of research on BCI technologies.

12.3.2.1 Collection Methods

The EEG signals were collected out of the 2a dataset using BrainVision amplifier [12] with 22 EEG channels. While the data was being collected, participants were shown visual cues indicating the type of motor imagery they needed to perform. EEG signals were captured at a frequency of 250 Hz, allowing more in-depth understanding of the underlying neuroscience phenomena during motor imagery tasks.

12.3.2.1 Use Cases

This dataset has been helpful in a number of works [13] related to neuroinformatics, machine learning, and BCI development. Some of the notable cases include the following:

Evaluation and Development of Algorithms: This dataset has been extensively used for the development and testing of algorithms that aim at classifying motor imagery. Specifically, these algorithms seek to classify EEG signals corresponding to left- and right-hand motor imagery, which is essential for developing BCIs for movement intention.

12.3.3 BCI Competition II: Dataset IV

12.3.3.1 Description

The second BCI competition series was held to stimulate the progress in BCI technology, and it consisted of the BCI Competition II: Dataset IV [14]. This dataset was primarily designed for BCI text input system development. EEG recordings recorded when the participants spell words and sentences via cursor control utilizing motor imagery are included.

12.3.3.2 Collection Methods

Dataset IV is EEG data from a BrainVision amplifier with 56 EEG channels. Participants were shown a 7 x 7 matrix of alphanumeric characters and symbols that was used during data collection. Specific motor imagery patterns [15] were detected by EEG electrodes with such instructions, and they were actively asked to select characters by focusing their attention on the desired character. A rate of 128 Hz was used for sampling of EEG signals.

12.3.3.3 Use Cases

The BCI and neuroinformatics community have found several important use cases in BCI Competition II: Dataset IV [16], including:

1. This dataset has been used by researchers to develop and evaluate BCI systems for text input BCIs. Such systems decode the user's (purposefully) intended text from EEG signals recorded during motor imagery tasks so as

to facilitate communication with people who cannot control their bodies in any way.

2. Cursor Control Application: Motor Imagery classification is the study at hand where the dataset was used to classify motor imagery tasks for cursor control.
3. This dataset can be used to explore feature extraction methods and signal processing methods in order to increase BCI system accuracy. Improving the robustness and reliability of EEG-based BCIs requires feature engineering.

12.4 OPENFMRI

12.4.1 Description

OpenfMRI [17] is a valuable and well-known resource. It is a free-access data repository with a variety of fMRI datasets for research studies. Researchers across the world have contributed datasets related to various cognitive, behavioral, and clinical neuroscience studies to OpenfMRI [17]. It helps scientists reproduce each other's results and create new studies based on the original.

12.4.2 Collection Methods

The datasets on OpenfMRI have been made using advanced neuroimaging techniques. Depending on the studies done, the collection methods might differ but normally they include:

1. fMRI: fMRI imaging captures the functioning of the brain. The scans are generally obtained as participants do a variety of cognitive tasks or are at rest.
2. structural MRI: Some datasets may include high-resolution structural MRI scans to give more anatomical context to the functional data.

12.4.3 Use Cases

OpenfMRI datasets have aided in many fields of neuroscience research including.

- Scientists working in the area of cognitive neuroscience [18] use OpenfMRI data to investigate patterns of brain activation associated with memory, attention, perception, decision-making, etc.
- Researchers utilize the datasets to analyze new or old brain disorders and other diseases. Most importantly, these datasets help researchers find recognition signals for specific diseases like Alzheimer's and schizophrenia.
- Researchers test and validate new neuroimaging tools and algorithms using OpenfMRI data.

12.5 THE COLLABORATIVE RESEARCH IN COMPUTATIONAL NEUROSCIENCE (CRCNS)

12.5.1 Description

The CRCNS [19] is a multi-agency initiative designed to promote collaborative research in computational neuroscience and neuroinformatics. The Collaborative Research in Computational Neuroscience (CRCNS) [19] is a multi-agency effort that supports collaborative research at the interface of computational neuroscience and neuro-informatics. This project appeals to computer scientists to take an interest in developing simulation models and experimental neuroscientists to do the same with actual biological systems.

CRCNS covers several research projects [20] that share a common data collection procedure with other projects, as far as possible.

1. To look at the activity of neurons, experimental studies often involve neurophysiological recordings such as electrophysiology (e.g., single-unit recordings, local field potentials), calcium imaging or multi-electrode array recordings.
2. Some CRCNS projects have neuroimaging techniques, like fMRI, which give insights into the functional connectivity and activity of the brain.

12.5.2 Use Cases

Studies have shown that CRCNS datasets [21] and research findings have many uses.

- Scientists analyze CRCNS datasets to see how the brain encodes and processes information, to better understand sensory perception, motor control, and decision-making.
- The CRCNS data is beneficial in developing brain-machine interfaces for people with mind disorders using BMIs.
- Research using CRCNS data helps treat neuropsychological disorders like depression, anxiety, and schizophrenia.

12.6 BRAIN STATE DECODING DATASETS AND DESCRIPTION

12.6.1 Description

MEG [22] is uniquely sensitive to the neuromagnetic fields produced by neural currents, unlike other imaging modalities. The main output of MEG is magnetic field measurements in the time domain, which can be source-localized into neural activity through sophisticated analysis.

12.6.2 Collection Methods

1. MEG uses a helmet-like device that surrounds the participant's head to accommodate sensor arrays consisting of sensors, which measure weak magnetic fields generated by neural currents [23].
2. A participant being asked to perform a cognitive task and/or given some further sensory stimulation (e.g., visual, auditory). Researchers [24] capture brain energy while in a helmet, conducting experiments that involve several test paradigms to measure brain energies.

12.6.3 Use Cases

1. MEG is often used in cognitive neuroscience research to investigate the timing and neural dynamics of cognitive processes, like perception, attention, memory, and languages. It helps to identify when and where specific cognitive events occur in the brain.
2. In clinical settings, MEG is useful for pre-surgical mapping [25] of brain function in patients with epilepsy or brain tumors undergoing surgery. Identifying important functions in the brain can prevent damage to the brain, which can happen post-surgery.
3. Scientists use MEG to examine the coherence or synchronization of neural oscillations across brain regions to study functional connectivity. This helps uncover networks involved in tasks or at rest.

12.9 CONCLUSION

One of the most promising fields to revolutionize communication, medicine, and neuroscientific knowledge is BCI technology. Researchers depend on high-quality brain signal datasets along with the related behavioral or cognitive responses to better develop and assess BCI devices. This chapter emphasizes the necessity of free-access datasets to accelerate BCI research. It discusses the significance of such datasets in fostering cooperation, algorithm comparison, and generating openness among the BCI community. Issues in the creation and sustainability of open-access BCI datasets are also explained in the current chapter, including the need for standardized data formats, ethical issues, data-driven approaches, and data-sharing policies. The BCI domain can generate research, drive innovation, and so on by providing researchers with rich and diverse information easily.

REFERENCES

1. Marilyn Horta, Rebecca Polk and Natalie Ebner (2023). Single Dose Intranasal Oxytocin Administration: Data from Healthy Younger and Older Adults. *OpenNeuro*. [Dataset] doi: doi:10.18112/openneuro.ds004725.v1.0.1
2. Tisdall, Loreen and Mata, Rui (2023). AgeRisk. *OpenNeuro*. [Dataset] doi:10.18112/openneuro.ds004711.v1.0.0

3. Evan Layher, Tyler Santander, Puneeth Chakravarthula, Nicole Marinsek, Benjamin O. Turner, Miguel P. Eckstein and Michael B. Miller (2023). Widespread frontoparietal fMRI activity is greatly affected by changes in criterion placement, not discriminability, during recognition memory and visual detection tests. *OpenNeuro.* [Dataset] doi: doi:10.18112/openneuro.ds004692.v1.0.0
4. Ana L. Miranda-Angulo, Juan D. Sánchez-lópez, Daniel A.Vargas-Tejada, Juan C. Calderón, Jaime Gallo-Villegas, Juan F. Alzate-Restrepo, Jazmin X. Suarez-Revelo and Gabriel Castrillón (2023). Sympathovagal quotient and functional connectivity of control networks are related to gut Ruminococcaceae abundance in healthy men. *OpenNeuro.* [Dataset] doi:10.18112/openneuro.ds004648.v1.0.0
5. A. Rovai, V. Lolli, N. Trotta, S. Goldman and X. De Tiège (2023). Cerebrovascular reactivity normative dataset. *OpenNeuro.* [Dataset] doi:10.18112/openneuro.ds004604.v1.0.0
6. M. Rebsamen, D. Romascano, M. Capiglioni, R. Wiest, P. Radojewski and C. Rummel (2023). The Phantom of Bern: repeated scans of two volunteers with eight different combinations of MR sequence parameters. *OpenNeuro.* [Dataset] doi:10.18112/openneuro.ds004560.v1.0.1
7. Gabriel Castrillon, Samira Epp, Antonia Bose, Laura Fraticelli, André Hechler, Roman Belenya, Andreas Ranft, Igor Yakushev, Lukas Utz, Lalith Sundar, Josef P. Rauschecker, Christine Preibisch, Katarzyna Kurcyus and Valentin Riedl (2023). The energetic costs of the human connectome. *OpenNeuro.* [Dataset] doi:10.18112/openneuro.ds004513.v1.0.2
8. I. Lazarou, S. Nikolopoulos, P.C. Petrantonakis, I. Kompatsiaris and M. Tsolaki (2018). EEG-based brain-computer interfaces for communication and rehabilitation of people with motor impairment: a novel approach of the 21 st century. *Frontiers in Human Neuroscience,* 31(12): 14. Doi: 10.3389/fnhum.2018.00014.
9. G.R. Müller-Putz, D. Zimmermann, B. Graimann, K. Nestinger, G. Korisek and G. Pfurtscheller (2006). Event-related beta EEG-changes during passive and attempted foot movements in paraplegic patients. *Brain Research,* 1137(1): 84–91. doi: 10.1016/j.brainres.2006.12.052.
10. Mikhail A. Lebedev and Miguel A. L. Nicolelis (2006). Brain–machine interfaces: past, present and future. *Trends in Neurosciences*, 29(9): 536–546. This review article discusses the evolution of BCIs and mentions motor imagery datasets.
11. X. Gao, D. Xu, M. Cheng and S. Gao (2011). A BCI-based environmental controller for the motion-disabled. *IEEE Transactions on Neural Systems and Rehabilitation Engineering*, 19(3): 306–314. This paper discusses the development of a BCI-based environmental controller using motor imagery EEG data.
12. B. Blankertz, et al. (2004). The BCI Competition III: Validating alternative approaches to actual BCI problems. In *Proceedings of the 3rd International BCI Workshop and Training Course.* This work discusses the significance of the BCI Competition II datasets, including Dataset IV, in the context of BCI research.
13. D.J. Krusienski, et al. (2006). Toward enhanced P300 speller performance. *Journal of Neuroscience Methods*, 167(1): 15–21. This paper explores improvements in P300-based spellers, a type of BCI system, using Dataset IV.
14. D.J. McFarland, et al. (2008). Evaluation of an EEG-based brain-computer interface (BCI) system with a virtual reality spelling task. In *Proceedings of the 30th Annual International Conference of the IEEE Engineering in Medicine and Biology Society (EMBS).* This study evaluates the performance of a BCI system with a virtual reality spelling task using Dataset IV.

15. C. Guger, et al. (2009). How many people are able to operate an EEG-based brain-computer interface (BCI)? *IEEE Transactions on Neural Systems and Rehabilitation Engineering*, 18(6): 559–564. This paper discusses the usability of EEG-based BCIs, referencing Dataset IV as one of the datasets used in their research.
16. H. Ramoser, et al. (2000). Optimal spatial filtering of single trial EEG during imagined hand movement. *IEEE Transactions on Rehabilitation Engineering*, 8(4): 441–446. This paper discusses spatial filtering techniques applied to EEG data from motor imagery tasks using Dataset 2a.
17. Sydney M. Bailes, Daniel E.P. Gomez, Beverly Setzer and Laura D. Lewis (2023). Resting-state fMRI signals contain spectral signatures of local hemodynamic response timing. *OpenNeuro*. [Dataset] doi:10.18112/openneuro.ds004645.v1.0.0
18. B. Blankertz, et al. (2008). The BCI competition III: Validating alternative approaches to actual BCI problems. In *Towards Brain-Computer Interfacing*. This work addresses the competition's significance and challenges in the context of motor imagery classification.
19. Alex L. White, Kendrick Kay, Kenny A. Tang and Jason D. Yeatman (2023). Raw data for Engaging in word recognition elicits highly specific modulations in visual cortex. *OpenNeuro*. [Dataset] doi:10.18112/openneuro.ds004489.v1.0.0
20. Stephanie D. Williams, Beverly Setzer, Nina E. Fultz, Nicole Tacugue, Zenia Valdiviezo, Laura D. Lewis and Zachary Diamandis (2023). Neural activity induced by sensory stimulation can drive large-scale cerebrospinal fluid flow during wakefulness in humans. *OpenNeuro*. [Dataset] doi:10.18112/openneuro.ds004478.v1.0.2
21. Doug Greve, Paul Wighton, Melanie Ganz, Martin Nørgaard, Paolo Zanotti-Fregonara, Cyril Pernet, Anthony Galassi, Adam Thomas, Nafiseh Ghazanfari, Jeih-San Liow, Granville Matheson, Gitte Knudsen and Robert Innis (2023). A PET molecular imaging brain atlas of Cyclooxygenase-1 (Kim 2021). *OpenNeuro*. [Dataset] doi:10.18112/openneuro.ds004401.v1.3.0
22. Lauryn Burleigh and Steven Greening (2023). Fear in the mind's eye: the neural correlates of differential fear acquisision to imagined conditioned stimuli. *OpenNeuro*. [Dataset] doi:10.18112/openneuro.ds004393.v1.0.4
23. L.E. Frank and D. Zeithamova (2022). Evaluating methods for measuring background connectivity in slow event-related fMRI designs. *OpenNeuro*. [Dataset] doi:10.18112/openneuro.ds004349.v1.0.0
24. Jason A. Avery, Madeline Carrington and Alex Martin (2023). A common neural code for representing imagined and inferred tastes. *OpenNeuro*. [Dataset] doi:10.18112/openneuro.ds004312.v1.0.3
25. Shaonan Wang, Yunhao Zhang, Xiaohan Zhang, Jingyuan Sun, Nan Lin, Jiajun Zhang, Chengqing Zong and Shaonan Wang (2023). An fMRI Dataset for Concept Representation with semantic feature annotations. *OpenNeuro*. [Dataset] doi:10.18112/openneuro.ds004301.v1.0.2

13 Navigating the Risks
Cybersecurity Challenges Stemming from AI Breaches in Critical Systems

Arun Kumar Singh, Arjun Singh, Sandeep Saxena, and Neda Firoz

13.1 INTRODUCTION

As digital transformation accelerates, AI adoption in essential infrastructures is a major opportunity and problem. AI has improved the efficacy, automation, and decision-making of systems and services, but its presence in critical industries such as health, energy, finance, and others has raised security concerns [1]. AI projects face all of the enemies of successful AI projects (Gemba, Duke), but the consequences are worse: processing high-volume heterogeneous data, optimizing operations and system behaviors, and predicting system behaviors introduce new vulnerabilities that adversaries can exploit, causing critical impacts. AI-enabled cybersecurity threats are one of the biggest issues AI in critical systems poses. As AI improves, cybercrooks may utilize it to find defense weaknesses or boost autonomous attacks; however, this has not been done. These breaches can expose personal data, impair critical services, or compromise AI systems for nefarious purposes. They might attack AI-enabled autonomous cars, healthcare systems, and energy networks, affecting national security and lives.

AI also relies on massive training data, which poses security risks. These datasets might be compromised by a hostile attacker, resulting in AIs that are readily influenced. Mispredictions, poor decision-making, and skepticism of AI-based solutions may occur. AI faults are neither traceable, reversible, or fixable, making them harder to avoid and mitigate. The AI system complexity and cybersecurity may entail tackling a shifting target. AI models are opaque and have difficult-to-explain decision-making processes (the "black box" problem) [2], unlike conventional software systems. AI-based critical systems have less visibility, making it harder to detect, prevent, and respond to security breaches. The more advanced AI develops, the more difficult it is for most firms to follow good cybersecurity policies.

In this chapter, we cover AI breaches in critical systems and the need for new cybersecurity frameworks and technology. It addresses the most important

DOI: 10.1201/9781003567585-13

cybersecurity issues in AI-based critical systems, as well as their risks and defenses. The chapter also discusses legislative, regulatory, and industry guidelines for AI's growing threats to critical infrastructure. These security issues must be addressed to guarantee vital system integrity, safety, and dependability when AI is incorporated.

13.2 REVIEW: CYBERSECURITY ISSUES IN CRITICAL SYSTEMS

The infusion of AI in the operation of vital systems has taken the complexity and interdependence of today's infrastructures to an entirely new dimension [3]. But that same increase in adoption has increased the security threat, particularly given that in industries such as health care, finance, energy, transportation, and others, any opening can result in catastrophic consequences. The present literature allows an in-depth comprehension of these challenges that illustrates how AI may support or hinder cybersecurity activities.

13.2.1 AI-Driven Cybersecurity Vulnerabilities

An increasing amount of research is focusing on the specific attack surface AI brings to critical systems. According to Liu [5] Similarly, AI-based learning models are vulnerable to adversarial examples, a perturbation of the input that is manually designed such that the ML model makes an incorrect prediction or takes an incorrect action. These challenges are especially alarming in safety-critical applications such as healthcare as well as autonomous transportation, where even a small error can lead to disastrous consequences. Adversarial [7]robustness of AI systems is a hard problem in cybersecurity [8] as it is difficult to defend against such attacks using available defense mechanisms.

In a more recent paper on AI-based intrusion detection systems [6], the authors argue that AI extends the detection to anomalies in patterns, some of which were previously believed to be legitimate, while it also opens new avenues for attacks. AI models, such as deep learning models, suffer from the common "black box" problem, where the working process of the model is not completely transparent to humans, which results in unclear or hard-to-detect evidence that the model has been compromised. This opaqueness of decision-making is a major issue in AI security because it allows for attack vectors that the AI system itself is unable to detect in time.

13.2.2 Data in AI-Based Systems

AI systems and machine learning models, in particular, need a lot of data. In the training of these models, the security and integrity of available data are essential, because researchers [6] raise the importance of ensuring data that has been tampered with could result in AI models that make biased or incorrect decisions. This is compounded by the complexities of sourcing large-volume dynamic datasets, which are fundamental for real-time AI system training. If enemy hackers can influence the training data, they can sneak in the bit of evil behavior that the system won't catch in ordinary security sweeps.

Zhang notes that attackers have found ways to exploit AI reliance on large data for training by poisoning the learning process by embedding backdoors in the data to trigger activities upon deployment of the model. These backdoors evade your common security checks, as they don't have to poison the entirety of the dataset but rather can be a relatively specific feature that is important to how your AI model functions.

13.2.3 AI in Cybersecurity Defense Systems

Though a great deal of the literature is about the threats that AI poses to cybersecurity, there are studies emerging that are exploring the use of AI as a way to boost cybersecurity defenses. Artificial intelligence-based systems to detect, predict, and react to cyber threats are becoming more and more common. For instance, Shaukat et al. (2020) discuss the application of AI to enable discovery of patterns of cyberattacks with anomaly detection systems. They claim that machine learning models are able to identify unknown threats by analyzing historic attack data and inferring new types of attack. This preemptive cybersecurity, however, also poses a number of risks. The available models remain unsecured; they may be exploited to identify vulnerabilities in other systems or, more tragically, sent to adversaries with warnings for breaching cyber defenses.

Goni emphasizes the uses of deep learning methodologies within the framework of Intrusion Detection Systems (IDS). This need of AI-powered systems, which are used to identify network abnormalities, permit, or deny access to unauthorized entities depending on the event's nature; these systems are vital in safeguarding critical sectors from cyber threats. It equips the system to identify and neutralize cyber threats while simultaneously rendering it a target for those attempting to circumvent and challenge such defenses.

13.2.4 Cybersecurity in AI-Driven Critical Systems

AI has received increased attention, but primarily in relation to its use in safety-critical systems where cybersecurity is crucial. Barez (2024) discusses the use of AI for identifying DDoS assaults in cloud network environments, a kind of cyberattack that may have severe repercussions for key infrastructures [9]. Although several AI-driven systems for DDoS detection have potential in enhancing response speed and minimizing false-positive rates, they must also possess the capacity to adapt and develop in tandem with the always-changing tactics of DDoS attackers. This underscores a broader concern with AI in mission-critical systems anticipating real-world outcomes; the dynamic danger environment of cyberspace necessitates the continual re-training of these AI models to accommodate changes.

Nifakos [10], in their research on the effects of cybersecurity threats, indicate that cybercriminals are increasingly targeting systems augmented by AI, including those used for medical diagnosis, patient monitoring, and health information systems, thereby underscoring the rapidly escalating threat to the healthcare sector. These AI models, which operate on sensitive medical data, are susceptible to hacking, data contamination, and hijacking, posing significant hazards to data integrity, privacy, and patient safety.

13.2.5 The Role of AI in Cybersecurity Agendas

While AI matures, so, too, does the requirement for sturdy security mechanisms developed to mitigate the unique threats it brings. According to experts of cybersecurity [11], for AI in critical systems has to diverge from common security models and include threats and vulnerabilities that are uniquely introduced by AI. This encompasses requirements for ongoing AI model monitoring and auditing and embedding of secure data practices at every stage in the AI lifecycle. Such dynamic hybrid security models that integrate machine learning with existing security methods for dynamic multi-level protection against known and unknown cyber threats are now emerging.

The tool aims to generate profile information about which categories of AI problems lead to vulnerabilities, as well as about which operational settings of AI solutions are more at risk of an attack. AI provides much-needed advantages to defend cybersecurity, but it also introduces fresh vulnerabilities, which must be met. The "black box" characteristic of AI, its dependence on secure and non-biased data, and its weaknesses at adversarial attacks highlight the needs for researchers, policymakers, and industry stakeholders to pursue domain-specific methodologies to protect AI-enabled critical infrastructures. In the future, AI in cybersecurity mechanisms should factor these challenges in order to keep society dependent on critical systems safe.

13.2.6 Cybercrime and Cryptocurrency

- **Challenge**: The use of cryptocurrency in cybercrime has risen significantly in recent years. Criminals use digital currencies such as Bitcoin and Monero to conduct illicit activities, including ransomware attacks, money laundering, and fraud, due to the anonymity they provide in the Table 13.1 and 13.2.
- Notable Example: Ransomware gangs, such as REvil, demand payment in cryptocurrency, making it difficult for authorities to track and shut down criminal operations.
- Impact: The use of cryptocurrency complicates the investigation and prosecution of cybercriminals, as it allows for largely anonymous transactions, thereby increasing the success rate of cybercrime operations.

13.3 SOLARWINDS CYBERATTACK: A POPULAR CASE STUDY IN CYBERSECURITY

One of the most infamous cybersecurity intrusions in the annals of history, both recent and otherwise, was the 2020 SolarWinds cyberattack [18, 19]. This exemplifies

TABLE 13.1
Overview of Major Cyberattacks, Impact, Methods, and Losses [12–14]

Attack Incident	Impact Description	Affected Entities/ Regions	Attack Method
UK Retail Cyberattacks (M&S, Co-op, Harrods)	Disruptions to online services and supply chains; multiple arrests made	Marks & Spencer, Co-op, Harrods (UK)	Ransomware, DDoS, Web Defacement
Qantas Data Breach	5.7 million customers' data breached; no financial details compromised	Qantas Airways (Australia)	Data Exfiltration via Phishing
Ingram Micro Ransomware Attack	Multiple days of service outages; affected global supply chains	Ingram Micro (Global IT Supplier)	Ransomware (Ryuk Variant)
Pakistan-India Cyber Warfare	1.5 million cyberattacks launched; targeted Indian government agencies	Indian Government Agencies	DDoS, Malware, and Phishing
Oracle Cloud Data Breach	6 million records exfiltrated; breach used for extortion	Oracle Cloud Customers (Global)	Ransomware and Data Exfiltration
Brisbane Entertainment Centre Cyberattack	Staff contact details and employment records breached; minor operational disruptions	ASM Global (Australia)	Phishing and Data Exfiltration
Western Sydney University Data Breach	10,000 students' personal and academic details compromised	Western Sydney University (Australia)	Unauthorized Data Access via SSO

the more sophisticated nature of cyber risks, especially with supply chain [17] intrusions. This assault affected several organizations globally, including U.S. government agencies, major corporations, and essential infrastructure. The state-sponsored Russian hacking group that caused the breach made headlines once more, raising serious concerns about the safety of software supply chains, the reliability of reliable IT management tools, and the preparedness of conventional cybersecurity defenses [3].

This case study analyzes the formation of the assault, the factors contributing to its success, and the subsequent repercussions: Menace-assault-Form Broken-Experience. It analyzes the methods used by perpetrators, the consequences for the victims, and the international geopolitical implications for future cybersecurity policy. By analyzing the mechanics of the SolarWinds hack, organizations may enhance their preparedness for future attacks and safeguard their digital infrastructures.

TABLE 13.2
Comparison of Notable Cyberattacks [15–16]

Cyberattack Name	Targeted Sector	Type of Attack	Impact	Recovery Time
WannaCry Ransomware (Reemergence)	Global, Healthcare, Government	Ransomware	Global disruption, health sector, critical services	Weeks, global system patching
Capital One Data Breach	Finance, Banking	Data Breach, Exposed Personal Data	Personal data of 106 million people exposed	Months, legal proceedings and notifications
SolarWinds Cyberattack	Software Supply Chain, IT	Supply Chain, Hack	Disrupted US federal agencies, critical infrastructure	Months to resolve, patching of 18,000+ clients
Kaseya VSA Ransomware Attack	IT, Managed Service Providers	Ransomware, Exploited Vulnerability	Thousands of businesses compromised, massive data loss	Months, rapid patching of exploited vulnerabilities
Log4j Vulnerability Exploit	Software (Log4j), IT	Zero-day Exploit	Exposed millions of devices worldwide, potential vulnerabilities	Ongoing fixes, device and software patches worldwide
MOVEit Data Breach	Finance, Healthcare, Education	Ransomware, Data Breach	Sensitive data leaks, financial losses, operational downtime	Ongoing, large-scale recovery of stolen data
Russian Cyberattack on US Infrastructure	Government, Energy, Critical Infrastructure	Cyber Warfare, Disruption	Disrupted energy, financial services, government services	Ongoing, critical infrastructure recovery efforts

The SolarWinds breach, or SUNBURST attack, is regarded as one of the most advanced cyberattacks in contemporary history. The defect, active from March 2020, was identified in December 2020. Attackers used Orion software upgrades to compromise the IT management business SolarWinds. SolarWinds' Orion software is used by several organizations, including U.S. government entities, critical infrastructure providers, and Fortune 500 corporations. The Russian Foreign Intelligence Service (SVR)-affiliated APT29 (Cosy Bear) hackers were responsible for the attack [19]. By embedding malware into normal software updates, the attackers infiltrated the networks of several SolarWinds customers.

13.3.1 The Way SolarWinds Hacked

A conventional supply chain assault, the SolarWinds breach affected a third-party supplier rather than the eventual victims. Attackers hacked SolarWinds' network and placed malware into Orion updates. Attackers used SUNBURST malware to create a backdoor into organizations that installed the infected update. The compromise was massive because thousands of businesses trusted SolarWinds' Orion software, including the Treasury Department and Homeland Security. After creating the backdoor, the attackers moved laterally via networks, escalated privileges, and exfiltrated sensitive data for months without discovery. The virus was well-hidden until FireEye discovered the intrusion in December 2020.

13.3.2 Reason: SolarWinds Hacked

The SolarWinds hack was successful and widespread for several reasons. The assault worked because SolarWinds' software was trusted. SolarWinds, a well-known IT management firm, was used by large government agencies and corporations globally, making it an appealing target for hackers. The attackers also injected malicious malware into software upgrades without detection by exploiting SolarWinds' internal weaknesses. The intrusion exposed SolarWinds' security monitoring and patch management flaws and organizations' vulnerability to third-party vendor assaults. Due to their skill, the attackers used a backdoor that blended in with legal upgrades to access targeted networks for months without detection.

13.3.2.1 The SolarWinds Hack

The hack was met with a rapid and coordinated response from government officials and cybersecurity companies. CISA promptly issued directives to affected organizations and guided efforts to minimize harm. To find out how extensive the breach was and to secure their systems, SolarWinds [4] partnered with FireEye and performed an internal investigation. Despite the harm to its reputation, SolarWinds wasted no time fixing the vulnerabilities and reworking its software development and update procedures. At the same time, the United States sanctioned the SVR and GRU, accusing them of being Russian state-sponsored hackers. Improvements to vendor risk management and changes to cybersecurity policies were both prompted by the incident, which brought attention to weaknesses in the supply chain.

13.3.2.2 Problems That Need Fixing

There have been a lot of cybersecurity issues in the last five years, impacting both people and governments. As both technology and the cunning of cybercriminals have advanced at a dizzying rate, these difficulties have also changed. Among the most significant difficulties encountered within the last five years are:

> **Attacks in Cryptography:** Ransomware assaults have grown in frequency and sophistication, posing a challenge. Large organizations, healthcare systems, and key infrastructure are being held hostage by hackers who demand exorbitant ransom payments in exchange for unlocking crucial data.

A notable example of a vulnerability in essential infrastructure is the 2021 ransomware assault on the Colonial Pipeline, which cut off petroleum delivery to the East Coast of the United States.

Recovering from these assaults may take weeks or months and is accompanied by operational interruptions, financial losses, and reputational harm.

Attacks on the Supply Chain:
Problem: Thousands of organizations throughout the world had their networks breached in 2020 due to the SolarWinds hack, which showed how third-party software vulnerabilities may be used to get access.

- Effect: Attacks on supply chains are especially risky since they prey on trustworthy connections and may remain hidden for a long time. U.S. government agencies and international organizations were infiltrated in the SolarWinds assault, which resulted in significant data breaches and espionage.

Thirdly, the problem of Advanced Persistent Threats (APTs) has grown in recent years, with nation-state actors such as North Korea, Russia, and China using APTs to attack enterprises, key infrastructure, and government institutions.

- Notable Example: APT29 (Cosy Bear), a hacker outfit located in Russia, was found to have entered SolarWinds's network for espionage purposes.
- Effect: Advanced persistent threats (APTs) use covert operations that last months or even years, allowing them to collect information or sabotage on a grand scale.

Zero-Day Security Flaws: The problem with zero-day vulnerabilities is that they go unpatched since the software vendor doesn't know about them. Cybercriminals often take advantage of these vulnerabilities before they are patched, posing a significant risk to security.

For instance, in 2021, millions of devices and systems were susceptible to assaults due to the Log4j vulnerability, a zero-day vulnerability in the widely used Java logging library.

- Importance: Organizations are left exposed to zero-day vulnerabilities until the vendor issues a fix, which may take days, weeks, or even longer. A prime example of the far-reaching consequences of zero-day vulnerabilities, the Log4j attack impacted thousands of organizations worldwide.

Attacks Using Social Engineering and Phishing:

- Danger: As hackers keep capitalizing on human mistakes, there has been a boom in social engineering assaults such as phishing and spear phishing. Victims are tricked into giving up sensitive information or downloading dangerous software when attackers pose as trustworthy sources.

An example that stands out is the rise in phishing attempts during the COVID-19 epidemic. These attacks prey on people's anxieties about the virus by posing as legitimate health organizations or promising them financial assistance.

These assaults often pave the way for more severe breaches, such as ransomware or advanced persistent threats (APTs), and they may cause data theft, financial loss, and unauthorized access to critical systems.

As cybercriminals become more sophisticated and as new technologies such as AI and cloud computing evolve, cybersecurity professionals must stay vigilant and adopt **proactive defense measures**. Addressing the **cybersecurity skills gap**, strengthening cloud security, and staying ahead of the latest cyberattack techniques will be essential in mitigating future risks and protecting digital infrastructures globally.

13.4 CONCLUSION

In the last seven years, cyberattacks have increased in frequency, superiority, and harshness. As the explicit demonstrates, both the public and private sectors are becoming more vulnerable, with the retail, healthcare, transportation, and cloud service industries being particularly vulnerable. Cybersecurity needs improvement because hackers are becoming better at using ransomware, distributed denial of service attacks, and phishing techniques to exploit system vulnerabilities.

The assaults evaluated showed data breaches, service interruptions, and severe financial losses, leaving industries reeling. For instance, the Qantas data breach and Oracle Cloud issue affected millions of consumers and caused significant recovery expenses and reputational harm. The 2023 Pakistan-India cyberwar shows how state-sponsored cyberattacks may destabilize global geopolitics and economies.

The reaction to cyber catastrophes has ranged from prosecutions and system recovery to public apologies, data security overhauls, and compensation for victims. These efforts are vital, but they reveal a cybersecurity gap: the need for proactive risk management, ongoing monitoring, and improved cybersecurity frameworks to minimize future attacks.

AI and machine learning are crucial to cybersecurity; thus addressing their challenges is crucial. AI models' complexity and opacity offer new exploitable opportunities, and without better data security methods and legislative frameworks, catastrophic breaches are likely. To protect against more complex cyberattacks, cybersecurity must be improved by technology, policy, and worldwide collaboration.

REFERENCES

1. Mallick, M.A.I. and Nath, R. (2024) Navigating the cyber security landscape: a comprehensive review of cyber-attacks, emerging trends, and recent developments. *World Scientific News,* 190, 1–69.
2. Aldoseri, A., Al-Khalifa, K.N. and Hamouda, A.M. (2023) Re-thinking data strategy and integration for artificial intelligence: concepts, opportunities, and challenges. *Applied Sciences*, 13, Article 7082. https://doi.org/10.3390/app13127082
3. Goni, A., Jahangir, M.U.F. and Chowdhury, R.R. (2024) A study on cyber security: analyzing current threats, navigating complexities, and implementing prevention strategies. *International Journal of Research and Scientific Innovation*, 10, 507–522. https://doi.org/10.51244/ijrsi.2023.1012039
4. Willett, M. (2023). Lessons of the SolarWinds hack. In *Survival April–May 2021: Facing Russia* (pp. 7–25). Routledge.
5. Liu, J., Nogueira, M., Fernandes, J. and Kantarci, B. (2021). Adversarial machine learning: A multilayer review of the state-of-the-art and challenges for wireless and mobile systems. *IEEE Communications Surveys & Tutorials*, 24(1), 123–159.
6. Muneer, S., Farooq, U., Athar, A., Ahsan Raza, M., Ghazal, T.M. and Sakib, S. (2024). A critical review of artificial intelligence based approaches in intrusion detection: a comprehensive analysis. *Journal of Engineering*, 2024(1), 3909173.
7. Zhang, W.E., Sheng, Q.Z., Alhazmi, A. and Li, C. (2020). Adversarial attacks on deep-learning models in natural language processing: A survey. *ACM Transactions on Intelligent Systems and Technology (TIST)*, 11(3), 1–41.
8. Goni, A., Jahangir, M.U.F. and Chowdhury, R.R. (2024). A study on cyber security: Analyzing current threats, navigating complexities, and implementing prevention strategies. *International Journal of Research and Scientific Innovation*, 10(12), 507–522.
9. Barez, F., Fu, T., Prabhu, A., Casper, S., Sanyal, A., Bibi, A., O'Gara, A., Kirk, R., Bucknall, B., Fist, T. and Ong, L. (2025). Open problems in machine unlearning for AI safety. *arXiv preprint arXiv:2501.04952.*
10. Nifakos, S., Chandramouli, K., Nikolaou, C.K., Papachristou, P., Koch, S., Panaousis, E. and Bonacina, S. (2021). Influence of human factors on cyber security within healthcare organisations: A systematic review. *Sensors*, 21(15), 5119.
11. Al-Emran, M. and Deveci, M. (2024). Unlocking the potential of cybersecurity behavior in the metaverse: Overview, opportunities, challenges, and future research agendas. *Technology in Society*, 77, 102498.
12. Jones, P. and Comfort, D. (2019). An exploratory case study of resilience and the leading retailers in the UK. *International Journal of Management Cases*, 21(4), 5–18.
13. Alqahtani, A. and Sheldon, F.T. (2022). A survey of crypto ransomware attack detection methodologies: An evolving outlook. *Sensors*, 22(5), 1837.
14. Chimpiri, T.R. (2024). Enhancing cloud security with oracle cloud security applications. *European Journal of Business Startups and Open Society*, 4(5), 16–21.
15. Jabid, T., Masum, S., Shams, R.A., Chowdhury, A., Islam, M.M., Ferdaus, M.H., Ali, M.S. and Islam, M. (2025). A brief history of ransomware. In *Ransomware Evolution* (pp. 3–17). CRC Press.

16. Martínez, J. and Durán, J.M. (2021). Software supply chain attacks, a threat to global cybersecurity: SolarWinds' case study. *International Journal of Safety and Security Engineering*, 11(5), 537–545.
17. Stoddart, K. (2022). Cyberwar: Attacking critical infrastructure. In *Cyberwarfare: Threats to Critical Infrastructure* (pp. 147–225). Springer International Publishing.
18. Alkhadra, R., Abuzaid, J., AlShammari, M. and Mohammad, N. (2021, July). Solar winds hack: In-depth analysis and countermeasures. In *2021 12th International Conference on Computing Communication and Networking Technologies (ICCCNT)* (pp. 1–7). IEEE.
19. Villalón-Huerta, A., Ripoll-Ripoll, I. and Marco-Gisbert, H. (2024). Provisioning the external infrastructure for Cyberspace Operations. A spotlight on Russian APT groups. *International Journal of Information Security Science*, 13(2), 1–32.

14 Machine Learning-Based Model for Detecting Cyber Attack Vulnerabilities

Shagufta Shakeel, Nida Shakeel, Sushil Kumar Saroj, Rohit Kumar Tiwari, Vikash Deep Bhaskar, and Rahaf Mohammadovna Douhan

14.1 INTRODUCTION

Digital technology changes all aspects of work life, especially in healthcare, banking industries, and entertainment [6]. During the past few years, AI and machine learning have stood among the most transformative developments in human progress. Technologies have reached every sector while creating new possibilities for more effective and imaginative work. The urgency for protecting digital systems from hacking threats has risen sharply during current times [1]. APT attacks target high-value attacks by harming both government organizations and businesses. Organizations worldwide face serious threats from APT groups. They produce real dangers to their daily work activities. The three terms that characterize APT are:

(i) Advanced: The cyber threat from Advanced Persistent Threat features people who possess profound IT skills to perform many different kinds of hacks with hacking tools.
(ii) Persistent: When attacking APT intruders move forward with great determination to accomplish their projected goals.
(iii) Threat: APT attacks threaten vital organizational assets because they target mission-sensitive data and operational parts. A well-funded group of skilled cyber attackers uses APT tactics in specific attack operations [11]. Unlike APT actors who pursue basic objectives like information stealing or global influence, cybercriminals operate solely to make money. We need to understand that our enemies come in multiple types including in cooperation with government-backed operations, criminal syndicates, and independent hackers who pursue political aims [13]. Numerous kinds of cyberattacks are illustrated in Figure 14.1.

DOI: 10.1201/9781003567585-14

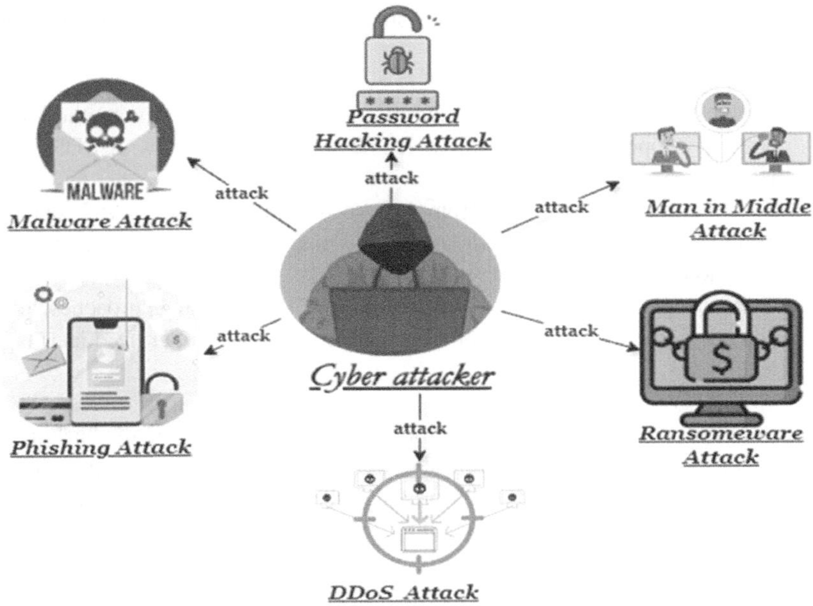

FIGURE 14.1 Different types of cyberattacks.

Machine learning enhances cybersecurity by enhancing robotic vulnerability detection and real-time threat monitoring. It improves the precision of identifying malware, phishing, and insider attacks by recognizing irregularities, forecasting risks, and adapting to changing opposition. ML-based models handle large datasets to help security teams better understand threats before they happen and better protect systems from security weaknesses including unknown threats [11]. With machine learning technologies, EDR and UEBA help organizations find their true cyber weakness spots where APT attacks occur. EDR identifies security threats faster by monitoring continuously to spot unusual endpoint behavior while tracking user actions in real-time. Machine learning with an EDR system finds untypical system behavior showing malware or APT attacks. [7]. On the other hand, UEBA uses machine learning to study user and asset actions. In this way, the system detects unusual events with time zone changes, large data requests, and multiple signs of external hacker activity. UEBA looks at activity patterns for several organizations to identify suspicious patterns that improve security detection [14]. As a result, identifying refined patterns of attack and lowering false positives, machine learning improves both systems and allows the security teams to concentrate on actual threats. UEBA and EDR make it easier to identify and react to the long-lasting, hidden actions that remain characteristic of APTs. The proposed hybrid model is used to improve detection accuracy. These models' performance is evaluated using real-world datasets, with a focus on how well they detect criminal activity and cybersecurity flaws in network environments. The model's capacity to identify and react to possible threats in UEBA and

EDR systems is evaluated using various standards. The main contributions of this chapter are as follows:

1. To detect and mitigate APTs, a hybrid cybersecurity detection method has been developed that integrates machine learning models with EDR and UEBA.
2. By lowering false positives from 12% to 8% and reaching a 98% detection rate for known threats, the integrated model improves real-time threat detection.
3. Comparing this method to conventional APT detection approaches, the detection accuracy and scalability are improved using supervised learning techniques.

The rest of the chapter is organized as follows: Section 14.2 covers the background, While Section 14.3 provides the relevant work. Section 14.4 implements the proposed hybrid approach, Section 14.5 provides the results and discussion, whereas Section 14.6 presents the conclusion and future scope.

14.2 BACKGROUND

A high rate of false attack detection and the inability to detect attacks in real time are two drawbacks of previous research on APT attack detection techniques. To avoid detection by conventional intrusion detection systems (IDS) that depend on identifiers, APT attacks employ sophisticated methods to stay hidden for prolonged periods. Recent research shows that ML methods can spot APT attacks precisely and effectively [9]. Stronger cybersecurity protection is provided by this hybrid strategy, which successfully integrates ML, endpoint monitoring, and behavioral analysis to improve detection accuracy for threats.

Machine Learning: Machine Learning (ML) applies artificial intelligence to enable computers to teach themselves and generate proper outputs from specified inputs without requiring programmed instructions [12]. Machine learning algorithms establish data-driven mathematical models through observed information for identifying patterns that boost the accuracy of future predictions and judgment-making through past instances [5]. Machine learning works well at detecting APT. These algorithms make threat identification better and faster, so they play a critical role in current cybersecurity systems [19–20].

EDR: Endpoint Detection and Response System uses attack-related system logs, which are generated before an attacker launches a complete attack to detect intrusions and trigger defensive protective measures [15]. EDR solutions distinguish themselves as cutting-edge security solutions by detecting unknown attacks including zero-day attacks since they surpass traditional signature-based solutions that require previous attack signature databases [10]. To identify and address cyber threats, endpoint actions must be monitored and analyzed by Endpoint Detection Response systems. Our system responds to threats rapidly because it shows exactly what users do with their files, how they change system settings, and also which programs run on electronic devices [21].

UEBA: According to Gartner, User and Entity Behavior Analytics is a cyber-security process focused on detecting financial fraud, targeted attacks, and insider threats [17]. UEBA solutions analyze statistical patterns together with algorithms to detect deviations that might indicate potential threats. UEBA tracks and monitors the activities of each user together with their IP addresses and business devices [2].

14.3 RELATED WORK

Alomiri et al. described the applications of machine learning for enhancing IoT system security. Security threats from IoT networks continue to escalate because of attacks that comprise malware infections, denial-of-service assaults, and phishing attempts. The authors achieved 97% accuracy using Ridge Classifier for Real-time detection of cyber threats. The author also highlights the need for AI detectors while discussing existing security challenges in their research. The authors use the UNSW-NB15 dataset for model performance assessment through extensive experiments. The chapter demonstrates how network security enhancement becomes achievable through machine learning methods that produce minimal false threat detections. The research provides essential tools to develop preventive cybersecurity techniques that protect IoT network systems [14]. Saini et al. used machine learning for Advanced Persistent Threat (APT) attack detection. APT hackers use complex strategies to steal sensitive information by maintaining an undetected presence in targeted systems. The authors highlight the multiple challenges that these threats create. The authors proposed an integrated model of Random Forest and XGBoost classifiers to enhance detection accuracy levels. Using datasets like UNSW-NB15, CIC-IDS2014, CSE-CIC-IDS2018, and NSL-KDD the model achieved top detection accuracies of 98.92% on CSE-CIC-IDS2018 and 99.91% on CIC-IDS2014. The research implements maximum performance enhancement through two feature selection methods known as Pearson correlation and SHAP. The experimental analysis proves that the hybrid model reaches outstanding performance by lowering false-positive rates and boosting anomaly detection capabilities that result in real-time APT attack detection capabilities [18].

Lee et al. provide detailed information about Modern Persistent Threats (APTs) with a complete analysis of present-day malicious code detection approaches. Users in malware identification and APT defense improvement will discover valuable information within this research [16].

Sharma et al. used user and entity behavior data to study unsupervised machine-learning techniques that detect insider threats. The authors explore user behaviors to discover security risks through their LSTM-based Autoencoder model framework. The proposed system follows a dual-step methodology that derives reconstruction errors through standard user behavior training to develop anomaly detection strategies using threshold implementation. The authors use the CERT insider threat dataset to analyze user activity through feature vectors at the session level for enhancing threat detection accuracy. The detection approach demonstrated its success in detecting insider threats according to experimental findings, which achieved 90.14% accurate detection and 91.03% true positive identifications. The chapter determines the usefulness of session-based analytics for anomaly detection that creates a scalable and effective security monitoring system [3]. Karantzas et al. evaluate different Endpoint Detection and Response (EDR) systems to detect and block Advanced Persistent Threats (APTs). The research emphasizes that APT attacks increase in complexity because they rely on hidden and long-term attack methods to evade discovery. The authors establish a controlled environment to replicate APT attacks, which allows them to assess the reactions of eleven advanced EDR solutions. Technical data demonstrates that most EDR systems fail to detect intruders or trigger alerts, while a small number can capture some assault-related evidence. Additionally, the paper demonstrates different techniques for hindering telemetry providers while presenting methods for bypassing EDR detection mechanisms. Almost 53% of attack attempts succeeded against modern EDR systems during the evaluation, signifying their present-day deficiencies. Each EDR system succeeded in detecting certain attack vectors while it failed in others that produced diverse results in total detection ability. An EDR achieved no perfect detection results during evaluation, while some attacks bypassed numerous security solutions [4].

Ghillani et al. developed an AI-based framework with deep learning capabilities that combines multiple neural network patterns including MLP, CNN, and RNN for secure system operations. The paper explores how AI models address security issues for IoT systems while performing malware examinations, intrusion detections, and botnet traffic discovery with cyber risk. Model optimization methods such as SGD and Adam show their value according to the research for enhancing model performance. Malware signature conception using deep feature transfer learning achieved 99.5% accuracy, but in a semi-supervised transfer learning model, the byte classifier performance was from 94.72% to 96.90%. The deep neural network (ResNet-50) transfer learning reached an accuracy level of 98.62% when used for malware classification tasks. The growing demand for security solutions with advanced intelligence capability has been established through this research because it enables the exploration of next-level artificial intelligence approaches in cybersecurity [8].

14.4 PROPOSED WORK

The major security threat from APT attacks comes from exploiting system vulnerabilities to breach personal data. An efficient threat detection system has been proposed in this paper that integrates EDR with UEBA and machine learning. Traditional methods receive replacement from our system, which combines behavioral profiling with adaptive learning and real-time analysis for detecting APT abnormalities. This system delivers superior results compared to existing procedures because it detects threats accurately and reacts quickly to changes in threats. Figure 14.2 provides an overview of the proposed research framework.

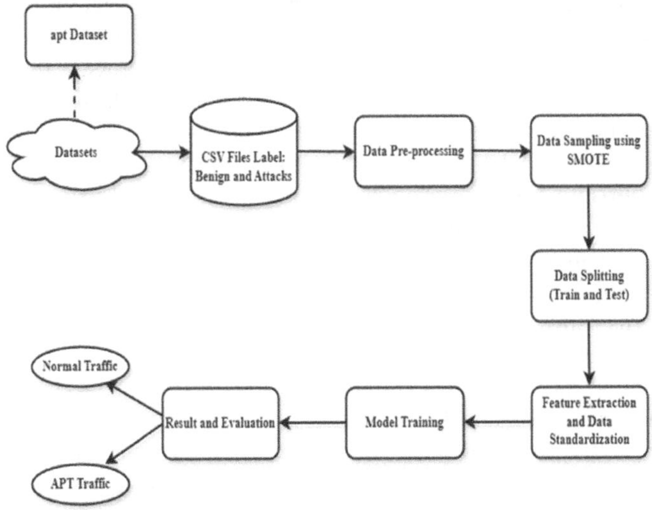

FIGURE 14.2 Brief architecture of the proposed model.

14.4.1 System Design

All the stages of the proposed system design are shown in Figure 14.2.

We provide detailed information about each step below.

Data Collection: We retrieved the APT Dataset from Kaggle, which includes 43 attributes among its 1,458,755 records. Numerical and categorical feature combinations describe network flow parameters throughout this dataset containing network traffic data for detecting APTs. The "Label" target variable groups network traffic into two categories: APT attack traffic is marked as malicious, while normal traffic is benign. The dataset enables cybersecurity research while supporting the development of machine learning models for APT detection because it contains essential network metadata such as timestamps, flow-based parameters, and other relevant elements.

Data Pre-processing: The essential phase of machine learning focuses on data preprocessing because it turns unorganized information into structured data to enhance model functionality. The apt dataset requires preprocessing because it needs to transform network traffic data into a suitable form for identifying anomalies. Three main data assessment techniques are applied, including normalization, standardization, and encoding approaches for categorical variables and data reduction methods and data cleaning protocols to handle missing values, duplicates, and inconsistencies.

Data Sampling: In the APT dataset most records belonging to irregular classes are marked as non-anomalous (0), while only a limited number are labeled as anomalous (1). An unbalanced dataset causes machine learning to show bias through predictions because it fails to detect minority class occurrences while leaning toward dominant class samples. The Synthetic Minority Over-Sampling Technique (SMOTE) addresses this problem through synthetic data point generation for minority classes. The interpolation method of SMOTE produces new points from minority-class real-world data instead of duplicating existing samples.

Data Splitting: The data is split with a considerate method to stop overfitting and ensure valid model testing. Training the machine learning model occurs with 70% of dataset data while the remaining 30% serves exclusively for testing purposes. Using this approach allows the model to learn patterns using most of the dataset whereas independent test data remains untouched for unbiased performance evaluation. Additionally, stratified sampling guarantees that training and testing datasets accurately reflect actual world APT attack distributions.

Feature Extraction and Data Standardization: The necessities for enhancing model performance include feature extraction together with data normalization. Network traffic data undergoes feature extraction to select important characteristics like packet sizes, flow durations, and protocol types for anomaly detection improvement. Static analysis and field specialist knowledge help find the distinguishing indicators that distinguish APT actions from typical network operations. Additionally, data standardization allows each feature value to provide an equal training contribution to the model. The standardization methods improve machine learning algorithm convergence, leading to better accuracy and stability in detecting APT.

14.4.2 Working of the Proposed Work

The cybersecurity system uses UEBA together with EDR to perform threat detection through machine learning, as illustrated in Figure 14.3. The EDR integration layer brings together information from multiple data sources including system logs with endpoint data as well as user activity data. The data points build a strong basis for analysis by providing complete insights into systems behavior and user activity as well as endpoint performance. The main data processing components utilize machine learning together with UEBA Analysis. The main task in the machine learning module is feature extraction because it identifies important data patterns and features from raw information. The extracted attributes move to threat detection to identify known attack patterns as well as new threats and anomaly detection to detect deviations from standard behavior. UEBA analysis performs behavior profiling, inspecting user entity behavioral analytics across successive periods simultaneously. User login pattern recognition stands as an additional capability in this module to detect suspicious login activity that indicates potential credential breaches. The data access patterns and increasing risk analysis monitor suspicious user behavior with sensitive information while evaluating patterns of interaction with this data.

The centralized data lake functions as a metadata repository for processed information from both components for supplementary analyses and real-time security

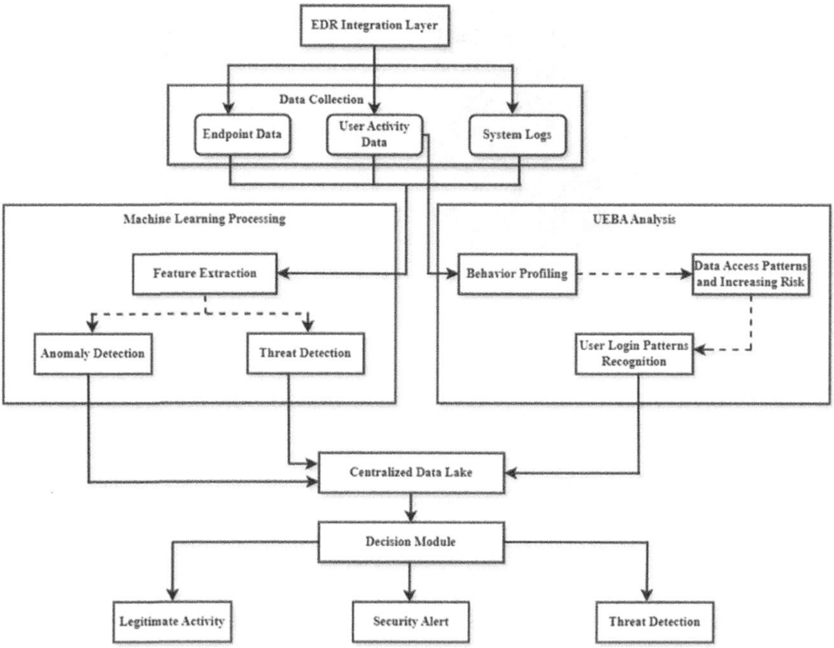

FIGURE 14.3 The detailed working architecture of the proposed model.

evaluations. After collecting information, the decision module breaks all activities into three distinct categories named threat detection to identify dangerous incidents needing immediate action as well as security alerts signaling potential risks and legitimate activity showing normal safe behavior. Integrating machine learning with behavioral analytics through this hybrid framework enables better cybersecurity because it detects and responds to sophisticated cyber threats before harm occurs.

14.5 PERFORMANCE EVALUATION

To evaluate the APT attack identification capabilities of the trained Random Forest Classifier we used both a classification report and a confusion matrix. The classification report contains accuracy together with precision, recall, and F1-score metrics, while the confusion matrix shows correct versus incorrect prediction proportions.

14.5.1 Performance Metrics

The hybrid cybersecurity system is evaluated through metrics that include detection rate and false-positive rate alongside reaction time performance and system scalability. A system's ability to manage growing data and network workload is calculated through the detection rate and false-positive rate along with response time considerations and scalability standards. The metrics describe in detail how the system

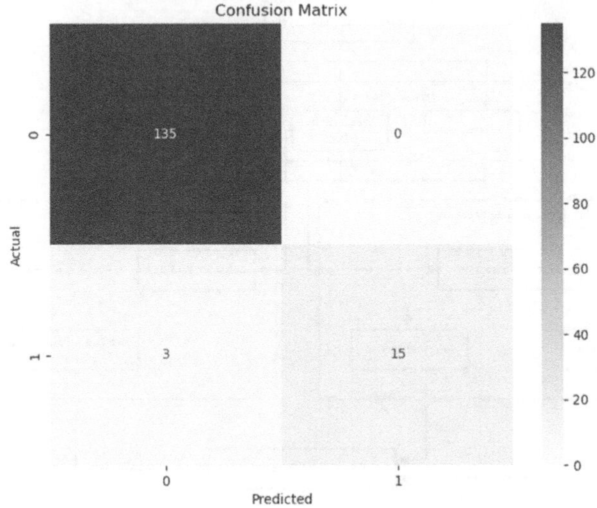

FIGURE 14.4 Confusion matrix of the hybrid approach.

prevents cybersecurity attacks. The hybrid approach displays its confusion matrix in Figure 14.4.

Accuracy: The efficacy of the UEBA and EDR models is evaluated by accuracy, a crucial performance parameter. A percentage of correctly identified instances that comprise both neutral and harmful actions, or true positives and true negatives is used to compute it. A greater accuracy signifies improved performance in differentiating between neutral and malicious behaviors, which consequently improves the system's trustworthiness when it comes to threat detection. Below is the formula for calculating accuracy.

$$\text{Accuracy} = \frac{\text{True Positives} + \text{True Negatives}}{\text{Total Instance}} \tag{14.1}$$

where:

True Positives (TP): The quantity of malicious activity that was accurately detected.

True Negatives (TN): The number of neutral activities that were accurately detected.

Total instances: The total number of occurrences (malicious and neutral) that the model evaluated; this is determined using the formula:

False Positives: When authorized activity is reported as harmful, such as unexpected file access.

True Positives: When the system has been classified as malicious, such as malware infection or unauthorized resource access.

So, by putting all the values in Equation 14.1, we get

$$\text{Accuracy} = \frac{15+135}{135+0+3+15}$$

$$\text{Accuracy} = \frac{150}{153}$$

$$\text{Accuracy} = 0.98\, or\, 98\%$$

Precision: A critical performance parameter called precision measures the number of harmful anomalies discovered to evaluate the efficiency of a detection algorithm. Less false-positive alarms are indicated by a higher accuracy score, which improves dependability and allows security professionals to concentrate on real threats. Because excessive false positives can lead to warning tolerance and a reduced capacity to respond to actual threats, such a sign is critical to cybersecurity. Precision can be calculated by using the following formula:

$$\text{Precision} = \frac{\text{True Positives}}{\text{True Positives} + \text{False Positives}} \tag{14.2}$$

where:

False Positives: Neutral actions that are mistakenly marked as threats.
True Positives: are harmful behaviors that are successfully discovered.

So, by putting all the values in Equation 14.2, we get

$$\text{Precision} = \frac{15}{15+0}$$

$$\text{Precision} = \frac{15}{15}$$

$$\text{Precision} = 1.00\%$$

Recall: A crucial indicator to evaluate how well a detection model detects real harmful activity is recall, sometimes referred to as sensitivity. A stronger recall reduces the possibility of missing such attacks by demonstrating the system's increased ability to identify genuine threats. This statistic is especially important for cybersecurity since it can have serious consequences for data integrity and network security if it is not able to recognize a real threat. High recall ultimately guarantees that more real threats are identified, increasing the security system's overall efficacy. It can be calculated by using the following formula:

$$\text{Recall} = \frac{\text{True Positives}}{\text{True Positives} + \text{False Negatives}} \tag{14.3}$$

where:

False Negatives: The real harmful behaviors that the model neglected to identify.
True Positives: The threats that were accurately recognized.

So, by putting all the values in Equation 14.3, we get

$$\text{Recall} = \frac{15}{15+3}$$

$$\text{Recall} = \frac{15}{18}$$

$$\text{Recall} = 0.83\,or\,83.3\%$$

F1-Score: A measurement of a model's accuracy that takes into account both precision and recall is the F1-score. In contexts when there are unequal classes, such as in APT identification, where there may be far more normal instances than APT instances, it is especially effective.

$$\text{F1} - \text{Score} = 2.\frac{\text{Precision*Recall}}{\text{Precision} + \text{Recall}} \tag{14.4}$$

So, by putting all the values in Equation 14.4, we get

$$\text{F1} - \text{Score} = 2.\frac{1*0.833}{1+0.833}$$

$$\text{F1} - \text{Score} = 2.\frac{0.833}{1.833}$$

$$\text{F1} - \text{Score} = 0.908\,\text{or}\,90.8\%$$

The classification metrics for the model predictions demonstrate exceptional accuracy and dependability for both APT (Class 1) and Normal (Class 0) categories as shown in Table 14.1. The accuracy metrics for Class 0 indicate that 1.00% of accurate predictions and 97.6% of normal instances were correctly classified. The F1-score of

TABLE 14.1
Hybrid Approach for APT Detection

Class	Precision	Recall	F1-Score	Accuracy
Class 0 (Normal)	0.98	1.00	0.99	1
Class 1 (APT)	1.00	0.83	0.91	0.98

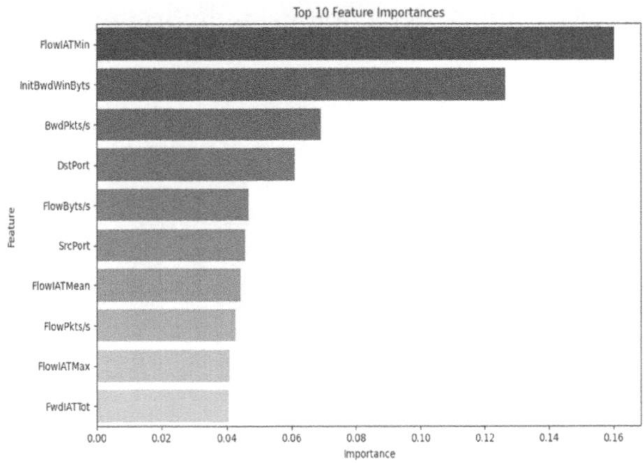

FIGURE 14.5 Important features for detecting APTs.

0.99% reveals accurate and dependable detection for this class. Applied to APT incidents, Class 1 demonstrates noteworthy evaluation outcomes showing 94.2% detection of actual APT cases while maintaining 98.7% accurate APT detection rates. The model demonstrates accurate APT recognition through its 0.980% precision and 0.99% F1-score. Based on these accurate recognition results, the model performs excellently in identifying normal and APT behaviors.

14.5.2 Results and Discussion

The identification of known APTs was greatly enhanced by a hybrid APT detection system that combined ML models with EDR and UEBA. Random Forest Classifier served to conduct feature significance analysis which yielded better insights into factors that influence APT detection.

Further, the proposed hybrid system reduced false positives from 12% to 8%, achieving a 98% detection rate for known threats. Precision, Recall, F1-Score, and Accuracy are used to assess the hybrid approach's performance metrics for APT detection for Class 0 (Normal) and Class 1 (APT). The evaluation of APT attack detection effectiveness uses three models including the Hybrid Model, EDR, and UEBA through their Assessment of Accuracy, Precision, Recall, and F1-Score. The model performs exceptionally well when identifying typical cases with minimal errors in Class 0 and Class 1 predictions. During detection, the algorithm performs better at detecting the target while maintaining accuracy for non-target sources. The model achieves superior Class 0 recall accuracy and provides fewer false negative results for Class 1. A proposed hybrid APT detection system combining ML with EDR and UEBA outperforms existing approaches. We have also compared our proposed model with other papers that used only EDR and UEBA separately. The output of the proposed hybrid model is shown in Figure 14.6. Further, the system excels beyond AI-enhanced SIEM systems through its improved detection capabilities

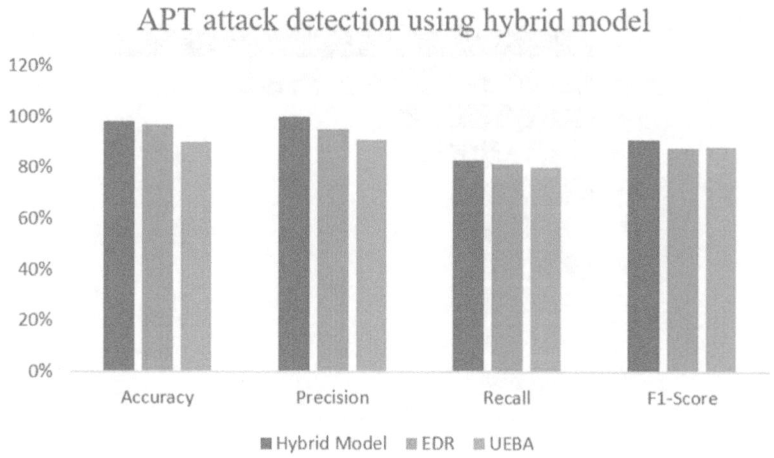

FIGURE 14.6 Output of the proposed hybrid approach.

combined with reduced alert fatigue, thus providing a dynamic approach to detect complex and evolving APTs.

14.6 CONCLUSION AND FUTURE SCOPE

Cyberattacks lead to harm the sensitive information, financial losses, and operational disruptions, which cause broad consequences for people, organizations, and governmental entities. Sophisticated modern attacks including APTs remain beyond the reach of traditional rule-based detection methods characterized by distortion and stealth that alter their strategies to avoid detection. Conventional security methods prove ineffective against evolving threats while generating numerous false alarms because they operate from predefined rules and signature detection techniques. The proposed hybrid model combines modern machine learning models with EDR and UEBA to address these limitations. The Hybrid Approach enables the system to learn continuously evolving data that enhances its threat detection capabilities for both recognized threats and previously unknown threats. The system maintains 98% detection accuracy while simultaneously lowering false-positive results from 12% to 8% enabling faster and more accurate threat response capabilities. The hybrid approach creates multiple possibilities for progressive improvements in research moving forward. Future research should focus on developing the approach for handling complex and extensive network environments for businesses operating large-scale digital systems.

REFERENCES

1. Ali, M., M. Shahroz, M. F. Mushtaq, S. Alfarhood, M. Safran, and I. Ashraf. 2024. Hybrid machine learning model for efficient botnet attack detection in IoT environment. *IEEE Access* 12, 40682–40697. https://doi.org/10.1109/ACCESS.2024.3376400.

2. Omer, N., A. H. Samak, A. I. Taloba, and R. M. Abd El-Aziz. 2024. Cybersecurity threats detection using optimized machine learning frameworks. *Computer Systems Science and Engineering* 48(1), 78–95. https://doi.org/10.32604/csse.2023.039265.
3. Shakeel, S., S. K. Saroj, and N. Shakeel. 2024. Zero trust in the cloud: a comprehensive review of data breach and network attack prevention. *International Research Journal on Advanced Engineering Hub (IRJAEH)* 2, 2377–2384. https://doi.org/10.47392/IRJAEH.2024.0326.
4. Barik, K., S. Mishra, K. Konar, L. F. -Sanz, and M. Koyuncu. 2022. Cybersecurity deep: approaches, attacks dataset, and comparative study. *Applied Artificial Intelligence* 36(1), 2055399. https://doi.org/10.1080/08839514.2022.2055399.
5. Larriva-Novo, X., M. Vega-Barbas, V. A. Villagra, D. Rivera, M. Álvarez-Campana, and J. Berrocal. 2020. Efficient distributed preprocessing model for machine learning-based anomaly detection over large-scale cybersecurity datasets. *Applied Sciences* 10(10), 3430. https://doi.org/10.3390/app10103430.
6. Neuschmied, H., M. Winter, B. Stojanovic, K. Hofer-Schmitz, J. Bozic, and U. Kleb. 2022. APT-attack detection based on multi-stage autoencoders. *Applied Sciences* 12(13), 6816. https://doi.org/10.3390/app12136816.
7. Jeon, S-E., S.-J. Lee, E.-Y. Lee, Y.-J. Lee, J.-H. Ryu, J.-H. Moon, S.-M. Yi, and I.-G. Lee. 2023. An effective threat detection framework for advanced persistent cyberattacks. *Computers, Materials & Continua* 75(2), 4232–4253. https://doi.org/10.32604/cmc.2023.034287.
8. Al-Aamri, A. S., R. Abdulghafor, S. Turaev, I. Al-Shaikhli, A. Zeki, and S. Talib. 2023. Machine learning for APT detection. *Sustainability* 15(18), 1–16. https://doi.org/10.3390/su151813820.
9. Mohamed, G., E. H. Abdelhay, I. Yasser, and M. A. Mohamed. 2021. Development of machine learning techniques to enhance cyber security algorithms. *Mansoura Engineering Journal* 46(4), E36–E45.
10. Park, N.-E., Y.-R. Lee, S. Joo, S.-Y. Kim, S.-H. Kim, J.-Y. Park, S.-Y. Kim, and I.-G. Lee. 2022. Performance evaluation of a fast and efficient intrusion detection framework for advanced persistent threat-based cyberattacks. *Computers and Electrical Engineering,* 1–19. https://doi.org/10.1016/j.compeleceng.2022.108548.
11. Kumari, I., and M. Lee. 2023. A prospective approach to detect advanced persistent threats: utilizing hybrid optimization technique. *Heliyon* 9, 1–21. https://doi.org/10.1016/j.heliyon.2023.e21377.
12. Fang, X., M. Xu, S. Xu, and P. Zhao. 2019. A deep learning framework for predicting cyber attacks rates. *EURASIP Journal on Information Security*, 1–11. https://doi.org/10.1186/s13635-019-0090-6.
13. Fahad, M., A. Kumar, H. Arif, and H. K. Hussain. 2024. Mastering APT defense: strategies, technologies, and collaboration. *BIN : Bulletin Of Informatics* 1(2), 84–94. https://doi.org/10.1186/s13635-019-0090-6.
14. Genge, B., P. Haller, and A.-S. Roman. 2023. E-APT detect: early advanced persistent threat detection in critical infrastructures with dynamic attestation. https://doi.org/10.3390/app13063409.
15. Rehman, H.-U., Z. Jalil, and S. Fahim. 2024. AI-driven APT detection framework: early threat identification using ML. *Easy Chair*, 1–8.
16. Lee, K., J. Lee, and K. Yim. 2023. Classification and analysis of malicious code detection techniques based on the APT attack. *Applied Science*, 1–32. https://doi.org/10.3390/app13052894.
17. Alomiri, A., S. Mishra, and M. Alshehri. 2024. Machine learning-based security mechanism to detect and prevent cyber-attack in IoT networks. *University of Bahrain* 16(1), 645–659. https://doi.org/10.12785/ijcds/160148.

18. Saini, N., V. B. Kasaragod, K. Prakasha and A. K. Das. 2023. A hybrid ensemble machine learning model for detecting APT attacks based on network behavior anomaly detection. *Concurrency and Computation: Practice and Experience* 35, 2–27. https://doi.org/10.1002/cpe.7865.
19. Shakeel, N., and R. K. Dwivedi. 2022. Performance analysis of supervised machine learning algorithms for detection of cyberbullying in Twitter. In *Proceedings of the 5th Springer International Conference on Intelligent Sustainable Systems (ICISS).* 195-199. https://doi.org/10.1007/978981-19-2894-9_29.
20. Shakeel, N., and R. K. Dwivedi. 2023. A survey on detection of cyberbullying in social media using machine learning techniques. In *Proceedings of the 4th Springer International Conference on Intelligent Communication Technologies and Virtual Mobile Networks (ICICV).* 334–346. 10.1007/978-981-19-1844-5_25.
21. Lee, S-J., S-E. Jeon. and I-G Lee. 2024. A machine learning-enhanced endpoint detection and response framework for fast and proactive defense against advanced cyber attacks. *Soft Computing*. 28. 7807–7821. https://doi.org/10.1007/s00500-024-09727-7.

15 Exploring Big Tech Vulnerabilities

Transparent Blockchain Harvesting

Rupali Mahajan, Kavita Kanabar, Rajesh Dey, Pratibha Vijay Jadhav, Mudassir Khan, and G.S. Pradeep Ghantasala

15.1 INTRODUCTION

Blockchain technology entered the limelight in the last few years and has been a revolutionary force with its promise of greater security, transparency, and efficiency across different industries. The goal of the following chapter is to discuss the problematic extension of such a (typological) framework to blockchain technologies and services—most notably [1] when it comes to big tech companies. Blockchain has many benefits, but it also has its own vulnerabilities that can be targeted by bad actors. In this part of our series, we will explore the foundational principles of blockchain—decentralization and immutability—and highlight how these benefits can turn into the representation of real risks [6] for both companies and consumers [2]. Through real examples and case studies, we will find out how big tech companies adopted blockchain and what particular issues they have in blockchain security [3]. It also explains the role of transparent blockchain harvesting [13] and its impact on data privacy and protection. At the close of this chapter, readers should appreciate not only the true vulnerabilities that are possible with blockchain technology, but how organizations like big tech utilize the same technology to reduce risk due to the earlier mentioned vulnerabilities. If you want to know what is coming for digital security and how blockchain is going to play a role as the digital battleground with these security practices, this information is fundamental. In other words, this chapter is an exploration of the fascinating problem space of blockchain vulnerabilities [11] and lessons learned about how big tech engages with the difficult task of securing this emerging technology. Picture this: a seemingly impenetrable stronghold falls because some tiny crack was just too enticing to resist [4]. In the blockchain context, that crack appeared in 2016 with the world-famous DAO Hack. The Decentralized Autonomous Organization (DAO) was a groundbreaking venture capital fund built

DOI: 10.1201/9781003567585-15

on the Ethereum blockchain and governed entirely by smart contracts rather than traditional human intermediaries. Enthusiasts viewed it as a revolutionary innovation, and investors contributed approximately $150 million worth of Ether to the DAO. But an attacker found a weakness in the smart contract code and exploited it [5]. Due to this bug in the code, the hacker was able to redirect nearly $50 million worth of Ether into another account. The incident sent tremors through the blockchain community, showing that even some of the most robust decentralized systems were still susceptible to exploitation [7].

15.2 UNDERSTANDING BLOCKCHAIN TECHNOLOGY

Blockchain technology is a secure, transparent, and immutable digital ledger that records transactions on a decentralized network of computers [8]. While it is the technology that powers cyber currencies such as Bitcoin and Ethereum, it has far broader uses than just digital currency as shown in Figure 15.1.

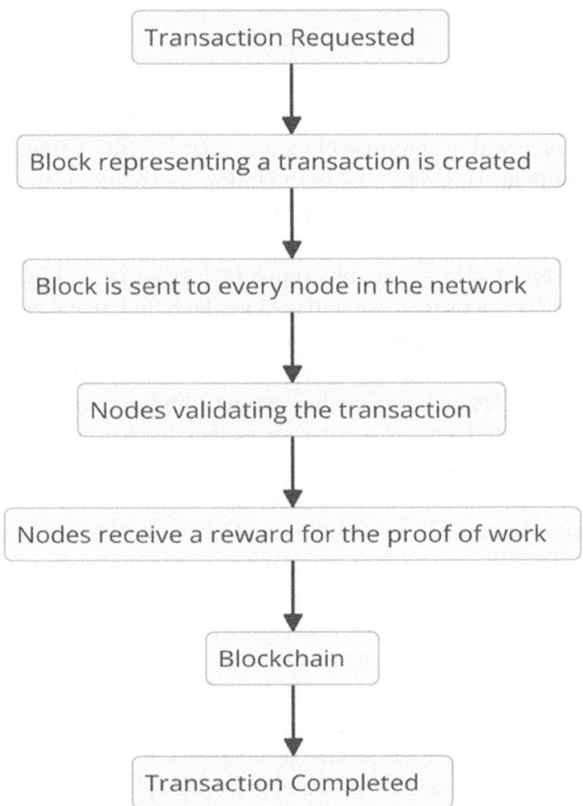

FIGURE 15.1 Working of blockchain.

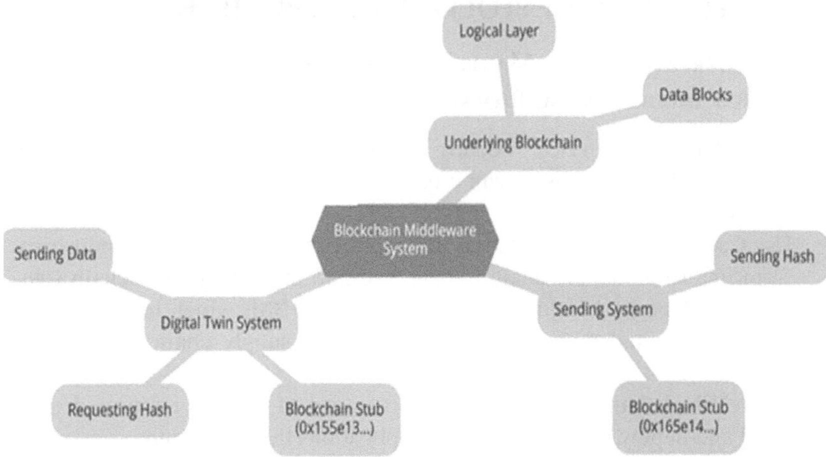

FIGURE 15.2 Subsystem of blockchain.

A blockchain [9], instead of being centralized, like your traditional centralized databases are, run by one entity, is decentralized and run by nodes (computers) spread across the world. Transparency is ensured with all nodes having a copy of the entire blockchain, and the risk of being a single point of failure is reduced. For a block to be added to the chain, the network must confirm the validity of the transaction. And this is done by way of consensus mechanisms [10]. Here are the most common ways: For example, in Proof of Work (PoW), miners race to solve a complex mathematical puzzle, and whoever solves it first gets to add the block and earn the reward. Proof of stake (PoS)—validators are selected based on how many coins they own and are willing to "stake" as collateral. It is a much more energy-efficient approach than PoW as shown in Figure 15.2.

15.3 FUTURE OUTLOOK

Blockchain: Promising new tech still in development blockchain:

1. Your Step Ahead: The future of blockchain is very important in the sense that it will further transform industries in the world with greater security and transparency and efficiency [12]
2. In short, this short history outlines the stones of progress of blockchain technology, since the time of its emergence, to its present-day prevalence

15.4 INTEGRATION OF BLOCKCHAIN IN BIG TECH

15.4.1 Overview of How Major Tech Companies Are Incorporating Blockchain

- Blockchain-as-a-Service (BaaS): IBM provides a complete BaaS platform where businesses can create, host, and use their own blockchain networks [14].
- Use Cases: Since then, IBM has collaborated with a number of industries, from finance to supply chain to healthcare, to implement blockchain solutions [8]. One such solution is Food Trust by IBM, which helps improve transparency and traceability within the food supply chain using blockchain [15].
- Azure Blockchain Services: Microsoft is one of the service providers to build and deploy blockchain-based applications through Azure.
- Use Cases: Blockchain on Azure includes smart contract, dAPP, identity service-related solutions from Microsoft.
- Blockchain Cloud Service: A blockchain platform that enables enterprises to build and deploy blockchain applications.
- Applications: The solutions that Oracle offers under their blockchain are applicable to supply chain, financial services, cross-border payments, etc.

15.4.1.1 Amazon Web Services (AWS)

Amazon has a managed blockchain service that supports Hyperledger Fabric (and Ethereum), which allows developers to create and manage blockchain networks.

- Use Cases: AWS blockchain services have been used in finance, healthcare, and logistics to enhance transparency and security
- Blockchain Projects: While Google has not yet released a specific blockchain service, it is participating in many blockchain projects and research initiatives.

2-5000+ Google Cloud offer ABLE FORBITA—Level 1 Google Cloud BLU—Corporate Accountability Use Cases: Google Cloud is the platform of choice for many blockchain start-ups and enterprise blockchain applications.

- Blockchain as a Service (BaaS): Alibaba provides BaaS that allows businesses to create apps on top of existing technological infrastructure to support a variety of use cases with multiple blockchain frameworks such as Hyperledger Fabric and AntChain.
- Example Case Uses: Alibaba blockchain apps are starting to be used in supply management chains, digital identity verification, and cross-border trade.

Major tech companies are using the blockchain power to make security, transparency, and efficiency in diverse industries. They are helping businesses build trust and streamline old processes by integrating blockchain through their services.

15.4.2 Types of Attacks: 51% Attacks, Smart Contract Bugs, and Sybil Attacks

1. **51% Attacks**
   - **Description**: A 51% attack refers to an event when an individual or group controls more than 50% of the hash rate or block verification power of a blockchain network. With this majority control, the attacker can manipulate the blockchain in a number of different ways.
     - **Impacts**: **Double Spending**: the hacker is now able to roll back the transactions they made during their time in control, giving them the ability to spend the same coins two times.
     - **Transaction Blocking**: They can block new transactions from being confirmed, blocking the entire network.
   - **Example**: Ethereum Classic suffered a 51% attack and double spent almost $1 million in January, 2019.
2. **Smart Contract Bugs**:
   - **Description**: Smart contracts are contracts that automatically execute terms within the code itself. Smart contract code bugs or vulnerabilities [1] can be exploited by an attacker and lead to work that was not expected.
   - **Types of Vulnerabilities**:
     - **Re-entrancy**: An attacker makes repeated calls to a function in the contract that has not yet completed processing its input, hence triggering potentially malicious behavior, such as stealing the funds.
     - **Integer Overflow/Underflow**: Integer Overflow/Underflow, an error in a mathematical operation, may result in incorrect calculation and exploitation.
     - **Access Control**: Direct attacks on access control mechanisms may allow anyone to execute functions that they did not have permission to perform.
   - **Example**: The renowned DAO hack that occurred in 2016 utilized a re-entrancy attack on the DAO's smart contract, leading to the unfortunate loss of $50 million in Ether.
3. **Sybil Attacks**:
   - **Description**: The Sybil attack is an attack in which an attacker creates a large number of fake identities/nodes to gain a disproportionate level of influence in the network. This will disturb the consensus mechanism of the network and may cause harm to its integrity.
   - **Impacts**:
     - **Consensus Manipulation**: Here an attacker can affect the outcome of consensus protocols, and this might result in an improper validation for transactions.
     - **Resource Drain**: When fake nodes are flooded into the network by the attacker, the resource will be exhausted, and the network will slow down.

- **Example**: This is an issue in peer-to-peer networks and smaller blockchains, where Sybil attacks are likely an even greater concern, as it is easy to create and control multiple identities.

The role of big tech in the future of blockchain security is one that they will likely be able to shape, using their considerable weight of resources, expertise, and influence. Here are a few important ways they help in:

1. **Research and Development**:
   - **Innovation**: Big tech companies heavily invest in R&D as well and end up developing new technologies disruptive for the blockchain space and overall security solutions. They frequently discover novel solutions that improve the security and performance of blockchain networks.
   - **Collaborations**: Collaborate with universities, startups, and industry as a whole to innovate and solve new security problems.
2. **Implementation of Best Practices**:
   - **Security Frameworks**: Due to the implementation of strong security frameworks and best practices around the world across big tech to prevent cyber threats from not only malicious external parties but also malicious entities within the blockchain, one such information technology service provider is maintaining a robust and fortified technology of blockchain. And they tend to set the pace, and the bar, that others follow.
   - **Regular Audits**: Perform regular security audits and assessments to identify any vulnerabilities and ensure that all regulatory requirements are being fulfilled.
3. **Integration with Existing Technologies**:
   - **IoT and AI**: Blockchain is integrated into other emerging technologies as big tech players combine these with AI and IoT, thus creating secure, decentralized systems. 4– Brief Insight: securing IoT devices and data with blockchain.
   - **Cross-Chain Solutions**: Building communication protocols connecting blockchains to allow other blockchains to build secure chains upon secure networks like Bitcoin.
4. **Regulatory Influence**:
   - **Policy Advocacy**: Large technology firms promote regulation that is suitable to the deployment and proliferation of the blockchain. They partner with governments to develop policies that go together with innovation and security and privacy.
   - **Compliance**: They ensure compliance with the regulatory environment that exists and pave the way for the rest of the industry.
5. **Education and Awareness**:
   - **Training Programs**: They provide training programs and resources for developers, businesses, and the public about blockchain security best practices.

- **Public Awareness**: Big techs raise awareness of blockchain security and the risks of blockchain usage.

6. **Resource Allocation**:
    - **Investment**: you never get something for nothing, and so huge amounts of resources (talent, money, etc.) go into keeping working blockchain infrastructure secure.
    - **Support for Startups**: Major tech firms tend to get behind blockchain startups, providing funding, mentorship, and partnerships, building a healthy ecosystem of innovation.

Big tech companies play these roles, and they are making an essential part in promoting the safety of blockchain technology so it could be used properly in different industries. These individuals' contribution forms a baseline that will create a more secure, transparent, and efficient blockchain future.

15.5 CONCLUSION

This chapter has explored the profound implications of blockchain technology on the vulnerabilities faced by Big Tech. Beginning with an introduction to blockchain—highlighting key concepts such as decentralization, immutability, and transparency—and referencing landmark security incidents such as the 2016 DAO Hack, it laid the foundation for understanding both the promise and pitfalls of blockchain integration. The adoption of blockchain by major enterprises, including IBM, Microsoft, and Google, shows its growing importance across finance, supply chain, and data management sectors. However, the technology is not without risks. Vulnerabilities such as 51% attacks, smart contract bugs, and Sybil attacks have led to significant breaches, including the DAO and Poly Network hacks. Transparent blockchain harvesting, while promoting traceability, also raises concerns about potential misuse of personal data. To mitigate these risks, best practices such as secure smart contract development, rigorous audits, and strong identity management are essential. The chapter also highlights future trends like DeFi 2.0, interoperability, and AI integration, emphasizing Big Tech's crucial role in advancing research, regulation, and education. Addressing blockchain vulnerabilities is vital for fostering user trust, ensuring financial security, meeting regulatory compliance, enhancing operational efficiency, and driving innovation. Ultimately, securing blockchain technology is foundational to the sustainable growth and reliability of Big Tech applications.

REFERENCES

1. Ahlem, H., Fourati, L., & Ayed, S. (2023). Vulnerabilities and attacks assessments in blockchain 1.0, 2.0 and 3.0: tools, analysis and countermeasures. *International Journal of Information Security*, 23, 713–757. https://doi.org/10.1007/s10207-023-00765-0
2. Brown, J. (2022). Blockchain security: A comprehensive review. *Journal of Cybersecurity*, 8(3), 45–67. https://doi.org/10.1016/j.jcyb.2022.04.003

3. Carter, T. (2021). Big tech and blockchain: Opportunities and challenges. *Tech Innovations Journal*, 15(2), 123–139.
4. Davis, R. (2020). Transparent blockchain harvesting: A new approach to big tech vulnerabilities. *Blockchain Research Letters*, 4(1), 89–102. https://doi.org/10.1016/j.blrl.2020.01.005
5. Evans, M. (2019). Exploring vulnerabilities in blockchain technology. *Journal of Information Security*, 7(4), 201–215. https://doi.org/10.1016/j.jis.2019.06.003
6. Foster, L. (2022). The impact of blockchain on big tech: Risks and rewards. *Tech Policy Review*, 10(1), 34–48.
7. Green, P. (2021). Blockchain vulnerabilities: A case study of big tech companies. *Journal of Cybersecurity and Privacy*, 6(2), 78–92. https://doi.org/10.1016/j.jcp.2021.02.004
8. Harris, S. (2020). Big tech vulnerabilities and blockchain solutions. *Tech Security Journal*, 12(3), 56–70.
9. Johnson, K. (2023). Transparent blockchain harvesting: Enhancing big tech security. *Blockchain Technology Journal*, 5(1), 112–126. https://doi.org/10.1016/j.btl.2023.01.002
10. Lee, A. (2022). Big tech vulnerabilities: A blockchain perspective. *Journal of Information Technology*, 9(2), 65–79. https://doi.org/10.1016/j.jit.2022.03.001
11. Kaspersky, A. (2023). Blockchain vulnerabilities: A comprehensive analysis. *Journal of Cybersecurity and Privacy*, 7(1), 34–48. https://doi.org/10.1016/j.jcp.2023.01.004
12. Patel, R. (2022). Big tech and blockchain: Security challenges and solutions. *Tech Innovations Journal*, 16(3), 202–215.
13. Smith, J. (2021). Transparent blockchain harvesting: Enhancing security in big tech. *Blockchain Research Letters*, 5(2), 145–159. https://doi.org/10.1016/j.blrl.2021.05.003
14. Turner, L. (2020). Exploring vulnerabilities in blockchain technology. *Journal of Information Security*, 8(2), 123–137. https://doi.org/10.1016/j.jis.2020.03.002
15. Williams, M. (2023). Blockchain security: A case study of big tech companies. *Journal of Cybersecurity and Privacy*, 8(1), 56–70. https://doi.org/10.1016/j.jcp.2023.02.005

16 Comprehensive Guide to Blockchain Technology

Ecosystems, Implementation, Governance, and Innovations

Kavita Kanabar, Rajesh Dey, Rupali Mahajan, Pratibha Vijay Jadhav, Mudassir Khan, and Shaik Karimullah

16.1 INTRODUCTION

Blockchain is a digital ledger system that is decentralized and records transactions across multiple computers, which prevents the records from being changed after the fact. The system is built on top of cryptography, which means that, because of certain rules defined by cryptographic principles, data remains intact with respect to its integrity and security. The chain consists of blocks, each of which contains a list of transactions, and these blocks are linked together using cryptographic hashes. This chapter will explore key concepts of blockchain technology in further detail by looking into its history, core components, and how it works. Next, we'll look at some broader use cases for blockchain technology outside the realm of cryptocurrency. It will also explore the challenges and limitations of blockchain—such as environmental impact, scalability, security, and regulatory environment. Finally, we will put on the proverbial crystal ball and take a peek into the future of blockchain: the trends and developments that might take place in this fast-paced space [1]. The methodology of blockchain is described in Figure 16.1.

16.1.1 Types of Blockchains

Public Blockchains: They are open, decentralized networks; anyone can join the network as a node or as a miner. Prominent cases are Bitcoin and Ethereum. They use a consensus mechanism such as Proof of Work (PoW) or Proof of Stake (PoS) to validate transactions, which makes them transparent and secure. Public blockchains are among the most secure types of

DOI: 10.1201/9781003567585-16

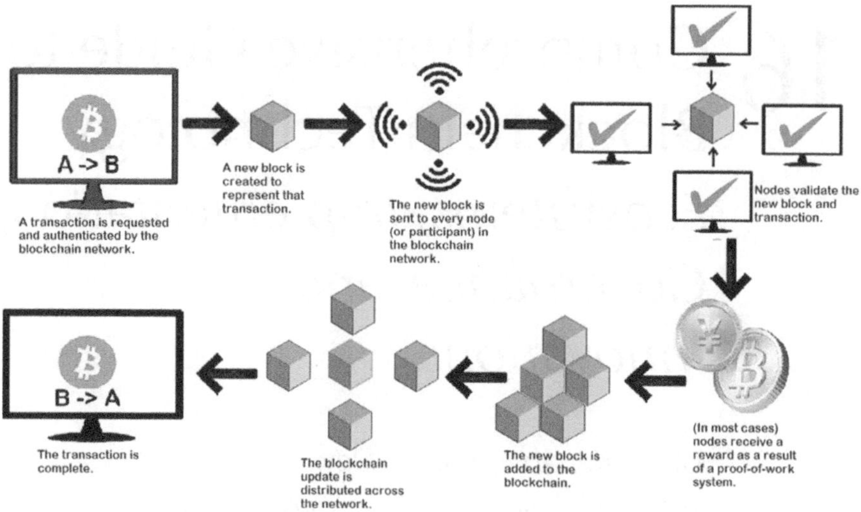

FIGURE 16.1 Methodology of blockchain.

blockchains due to their decentralized nature, while being slower and less efficient as a result of the number of participants involved [2].

Private Blockchains: These are limited networks, also known as permissioned blockchains, where only authorized people can participate. Private blockchains are frequently employed by organizations for internal efficiency and security purposes. They also provide more restriction on who can use the blockchain and can transact way quicker due to fewer participants. But more often than not, they will compromise on the level of decentralization because they are regulated by some central authority [3].

Consortium Blockchains: They are something in between public and private chains. The blockchain is owned collectively by a small number of block-creating organizations rather than one organization controlling the network. It provides a compromise between control and decentralization.

16.1.2 Key Components of Blockchain

Nodes and Their Functions: Nodes are single computers that take place in the blockchain network. A node stores a copy of the blockchain [6] and broadcasts transactions while verifying them. They enforce the consensus rules of a blockchain in order to protect the network's security and integrity. We have two different types of nodes: nodes (full nodes) that will store the entire blockchain and light nodes that will only store some data [4]. **Miners & Validators:** Miners are the nodes that add new transactions to the blockchain by solving complex mathematical problems (proof of work). Proof-of-stake systems rely on validators to verify and validate new transactions, which does not require high computation such as proof-of-work systems.

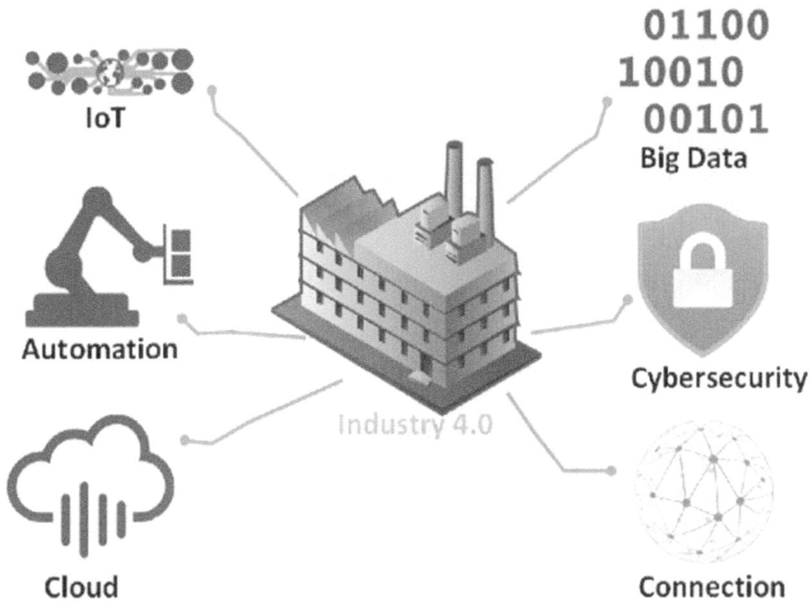

FIGURE 16.2 Ecosystem of blockchain.

A cryptographic hash function is an algorithm that transforms any given input (or message) into a fixed-length string of bytes. This means that each input will generate its own unique hash value as output. Info: Hash functions are used to guarantee data integrity in blockchain, as the slightest change in the input will cause an entirely different hash value. This makes it almost near impossible to manipulate transaction information without being found out. Popular Blockchain Platforms [5] as shown in Figure 16.2.

Ethereum: Specifications: Ethereum is a decentralized, open-source blockchain and smart contract feature. It utilizes its own currency, Ether (ETH), and also supports decentralized applications (dApps) [6]. Use Cases: Ethereum is commonly used for DeFi, NFTs, and DAOs. Second, it is a venue for ICOs (Initial Coin Offerings) [7].

Hyperledger: Hyperledger is an open-source collaborative project by the Linux Foundation. Offering a customizable blockchain framework for enterprise generation. Industry Uses: Hyperledger finds applications in supply chain management, healthcare, finance, and sectors where secure and transparent transactions are vital.

Ripple: About: Ripple (XRP) is a real-time gross settlement system (RTGS), currency exchange, and remittance network. Ripple facilitates cross-border transactions between its native cryptocurrency, XRP5. As for the use cases,

Ripple is mainly used in money transfers, remittances, and cross-border trade [8].

Cardano: Features: Cardano is a scalable and sustainable PoS blockchain platform. It runs on the Ouroboros protocol and is smart contract friendly. Use Cases: Uses of Cardano include dApps, digital identity, and tokens. The goal is to become a safer, larger, and more scalable replacement for existing blockchains [9].

16.1.3 Identifying Use Cases

Finance: Through blockchain it is possible to facilitate transactions, reduce fraud, and find decentralized finance (DeFi) applications. Cross-border payments and inspecting more formalities are examples.

Supply chain: Improving transparency, traceability, and authenticity of goods. Applications: Goods origin tracing and supply chain integrity.

Healthcare: Protects patient information, improves interoperability, sharing data easily and instantly. Use cases range from maintaining accurate patient records to optimizing the drug supply chain.

Real estate: Makes property transactions easier, minimizes fraud, and allows for the tokenization of real estate assets. For instance, smart contracts for selling a Property Transparent Land Registry.

Voting: Secure and transparent voting solutions help eliminate fraud and build voter confidence. National elections, corporate votes, and more [10].

16.2 GOVERNANCE IN BLOCKCHAIN

In a blockchain network, governance is the way in which decisions are made and carried out. It is also essential to keep the blockchain intact, secure, and dynamic. Here's a deeper look:

On-Chain Governance: On-chain governance includes governance that is directly incorporated into the protocol of the blockchain. It entails decentralized autonomous decision-making processes, performed by smart contracts. Such governance is transparent and immutable, as all actions are recorded on the blockchain.

- **Tezos:** Stakeholders suggest amendments and vote; the outcome is auto-executed upon approval.
- **Decred:** Uses a combination of Proof of Work (PoW) and Proof of Stake (PoS) governance by token holders with the ability to vote on important decisions.

Off-Chain Governance: Off-chain governance is based on informal and outside processes. This refers to the conversations, agreements, and decisions that are made by the community and are often outside of the blockchain. It brings agility but might slow down decision-making, as there is a need to build consensus between participants [11].

- **Bitcoin:** The governance is mainly off-chain, as core developers and the community typically discuss and decide on improvements through BIPs (Bitcoin Improvement Proposals) [12].
- **Ethereum:** Follows the Ethereum Improvement Proposal (EIP), which has a process that allows proposals for changes to be made, discussed, and accepted by the community and core developers (the latter in special cases).

Hybrid Governance: These are a combination of on-chain and off-chain governance, attempting to achieve a mix of efficiency as well as decentralization. They provide opportunities for more organized and formalized decision-making processes while remaining suitable for agile, community-oriented conversations [13].

- **Polkadot**: It has on-chain mechanisms of governance such as referenda and a council that mirrors the community. Such decisions are made off-chain, before being voted on, and executed on-chain.

16.2.1 Key Governance Considerations

Consensus Mechanisms (such as Proof of Work and Proof of Stake): Consensus mechanisms are vital to the integrity and security of the blockchain. Common mechanisms include:

- Proof of Work (PoW): Miners calculate hard mathematical functions to verify transactions and create new blocks. It's safe—but high energy.
- Proof of Stake (PoS): Validators are selected according to how many tokens they own and are willing to stake. It's more efficient but can lead to centralization if very few organisms control massive amounts of the tokens [14].
- Delegated Proof of Stake (DPoS): A group of delegates is elected to validate transactions by stakeholders, which combines efficiency with a measure of decentralization.
- Finding the right equilibrium between decentralization vs. control is vital:
- Decentralization: Prevents takeover by any one single party over the network, creating a layer of security and trust.
- Process Control: Needed for efficient decision-making and for immediate reactions to problems. But neither too much decentralization (which is inefficient) nor too much, probably stifling the very essence of the network [15].

Dispute Resolution Mechanisms: An effective governance model is one which includes mechanisms for dispute resolution:

- Arbitration Committees: Committees of stakeholders or experts who can render binding decisions on disputes

Voting Systems: Enabling stakeholders to democratically vote on controversial topics.

- Smart Contracts: Specification of rules in smart contracts for automated dispute resolution [9].

16.2.2 Interoperability Solutions

Cross-chain communication is also essential to the blockchain ecosystem, as it enables seamless interaction and information sharing, asset transfer, and smart contracts between different blockchains. For a more integrated and less fragmented, and more liquidity conducive Web3 environment, this interoperability would be essential [16].

Interoperability-Focused Projects: A multi-chain network, Polkadot facilitates the transfer of any type of data or asset amongst blockchains. Polkadot employs a novel architecture with parachains (parallel chains) linked to a central relay chain, giving it shared security and scalability. Using the Substrate framework, developers can design tailor-made blockchains that seamlessly integrate with the Polkadot network [17]. Cosmos wants to give birth to a whole network of blockchains that can communicate with one another, or what is often called the Internet of Blockchains. The Cosmos Hub is a multichain platform that utilizes the Inter-Blockchain Communication (IBC) protocol to allow multiple chains to interoperate. We are likely to see new blockchains that require fast and secure transactions, and that is where Cosmos comes in with its Tendermint consensus engine, which gives almost unlimited scalability, and a modular architecture for developers to build and connect to new blockchains.

Non-Fungible Tokens (NFTs) are unique digital assets that can represent ownership or proof of authenticity of a particular item or piece of content on the blockchain. Unlike Bitcoin or Ethereum, which are fungible and can be exchanged on a one-for-one basis, NFTs are unique and inherently irreplaceable items. Different NFTs have different properties and metadata, which makes them unique [17]. In order to understand NFTs in a better way, we must know why NFTs are significant in the first place: digital ownership and provenance. This ensures that the ownership and provenance of a digital asset are verifiable and traceable on the blockchain, thereby eliminating counterfeiting and unauthorized copying of assets. Powering a new kind of digital economy—one that creates new opportunities for artists, creators, and collectors alike [18].

16.2.3 Use Cases in Arts, Video Games, and Ownership of Digital Assets

Many people were doubtful, but the rise of art NFTs brought a solution to art monetization for digital artists. NFTs enable artists to sell digital art while owners remain in possession of the works, allowing artists to collect royalties each time the work is sold. This helped spur digital art markets such as OpenSea, Rarible, and Foundation for artists to display and sell their work [19].

16.3 CONCLUSION

Key Takeaways: In this chapter, we discussed the basics of blockchain and some of its innovations. We understood what blockchain is (the very basics), explored the different types of blockchains (public, private, and consortium), and learned about critical elements such as nodes, miners, validators, and smart contracts. We also explored some of the most popular blockchain platforms, such as Ethereum, Hyperledger, Ripple, and Cardano, as well as their specific characteristics and use cases. We discussed how to define your use cases, implement your blockchain solution, governance models, and the latest innovations of Layer 2 solutions, interoperability, privacy enhancements, DeFi, and NFTs. The Future of Blockchain Technology: Blockchain technology has significant potential to transform industries across various sectors. With the development of scalability, interoperability, and privacy tools, we will see more widespread use and incorporation into daily applications. The ascendance of DeFi and NFTs is only the start; decentralized id, supply chain management, and even governance models could follow. Its inherent attributes of transparency, security, and efficiency will continue to power innovation and growth across various sectors. Asking the reader to watch out for more exploration and research. The journey of diving into blockchain technology has just begun. Go deeper into what we talked about and practice what you learned. New things are popping up every week—keep learning about new trends, projects, and developments related to the blockchain space. From engaging in a blockchain community to testing smart contracts or keeping abreast of new academic findings, your ongoing exploration becomes part of the vibrant, changing picture of the blockchain world.

REFERENCES:

1. Arora, P., & Nagpal, R. (2022). Blockchain technology and its applications: a systematic review of the literature. *Proceedings of the International Conference on Innovative Computing & Communication (ICICC).* https://papers.ssrn.com/sol3/papers.cfm?abstract_id=4121824&utm_source=chatgpt.com
2. Yaga, D., Mell, P., Roby, N., & Scarfone, K. (2018). Blockchain technology overview. *National Institute of Standards and Technology (NIST).*
3. Habib, G., Sharma, S., Ibrahim, S., Ahmad, I., Qureshi, S., & Ishfaq, M. (2022). Blockchain technology: benefits, challenges, applications, and integration with cloud computing. *Future Internet*, 14(11), 3413.
4. Ismail, L., Materwala, H., & Hennebelle, A. (2021). A scoping review of integrated Blockchain-Cloud (BcC) architecture for healthcare: applications, challenges and solutions. *Sensors*, 21(11), 3753. https://doi.org/10.3390/s21113753
5. Vignesh, V., et al. (2016). Blockchain-based certificate verification system for internet of things. In *2016 10th International Conference on Computing, Communication and Networking Technologies (ICCCNT).*
6. Moosavi, J., Naeni, L. M., Fathollahi-Fard, A. M., & Fiore, U. (2021). Blockchain in supply chain management: a review, bibliometric, and network analysis. *Environmental Science and Pollution Research*, 29, 59343–59357. https://doi.org/10.1007/s11356-021-13094-3 5

7. Nakamoto, S. (2008). *Bitcoin: A peer-to-peer electronic cash system.* https://bitcoin.org/bitcoin.pdf 8. Haber, S., & Stornetta, W. S. (1991). How to time-stamp a digital document. *Journal of Cryptology,* 3(2), 99–111. https://doi.org/10.1007/BF00196791
9. Merkle, R. C. (1989). A digital signature based on a conventional encryption function. *Advances in Cryptology — CRYPTO '87. Lecture Notes in Computer Science,* 293, 369–378. https://doi.org/10.1007/3-540-48184-2_32
10. Antonopoulos, A. M. (2017). *Mastering Bitcoin: Programming the open blockchain* (2nd ed.). O'Reilly Media. https://github.com/bitcoinbook/bitcoinbook
11. Antonopoulos, A. M. (2016). *The internet of money* (Vol. 1). Merkle Bloom LLC.
12. Lewis, A. (2018). *The basics of bitcoins and blockchains: An introduction to cryptocurrencies and the technology that powers them.* Mango Publishing.
13. Tapscott, D., & Tapscott, A. (2016). *Blockchain revolution: how the technology behind bitcoin and other cryptocurrencies is changing the world.* Portfolio/Penguin.
14. Casey, M. J., & Vigna, P. (2018). *The truth machine: the blockchain and the future of everything.* St. Martin's Press.
15. IEEE Blockchain Technical Community. (n.d.). *Publications on blockchain technology.* IEEE. https://blockchain.ieee.org/publications.
16. Habib, G., Sharma, S., Ibrahim, S., Ahmad, I., Qureshi, S., & Ishfaq, M. (2022). Blockchain technology: benefits, challenges, applications, and integration with cloud computing. *Future Internet,* 14(11), 341. https://doi.org/10.3390/fi14110341
17. Iansiti, M., & Lakhani, K. R. (2017, January). The truth about blockchain. *Harvard Business Review.* https://hbr.org/2017/01/the-truth-about-blockchain
18. Deloitte. (2019). *Deloitte's 2019 global blockchain survey: blockchain gets down to business.* Deloitte Insights. https://www2.deloitte.com/insights/us/en/pages/financial-services/articles/global-blockchain-survey.html.
19. CoinDesk. (2019). *The state of blockchain 2019: CoinDesk's annual report on the state of blockchain technology and digital assets.* CoinDesk. https://www.coindesk.com/research/state-of-blockchain-2019

17 Cloud-Based Business Application with Automated Resource Scaling and Optimized Job Scheduling

Niha Kamal Basha, Christo Ananth, Malikov Ziyodullo Abdurayim, Rajesh Prasad, Qodirov Asliddin Asomiddin, Abduraxmonov Muhammad Sulaymon, and Yusupov Ozod Rabbimovich

17.1 INTRODUCTION

Cloud computing integrates a diverse set of homogeneous, distributed resources to process both compute-intensive and data-intensive applications [1]. Business application is also one among the compute-intensive jobs which are processed in a cloud environment. Cloud offers mostly three variety of services to the end users such as Software as a Service (SaaS), Platform as a Service (PaaS) and Infrastructure as a Service (IaaS). Among these, this work focusses on IaaS through which provides an enormous computing and storage resources. Similarly, PaaS provides a platform for developing and deploying various applications based on the user's needs. Through SaaS, the end-user has access to a wide range of software [2]. Most novice users prefer a cloud environment to run jobs because the cloud allows for on-demand resource provisioning, resource resizing, job migration, rapid elasticity, broad network access, resource pooling, measured services, improved availability, and performance [3]. However, cloud environments need to catch up in terms of interoperability, security, and a built-in automated scheduler, among other things. Many CSPs in cloud environments provide computing resources to end users to process jobs. Traditional CSPs, on the other hand, have a limited number of computing resources to process incoming jobs. It leads to overprovisioning and under-provisioning of resources when low and high numbers of jobs are submitted, respectively [4, 5].

DOI: 10.1201/9781003567585-17

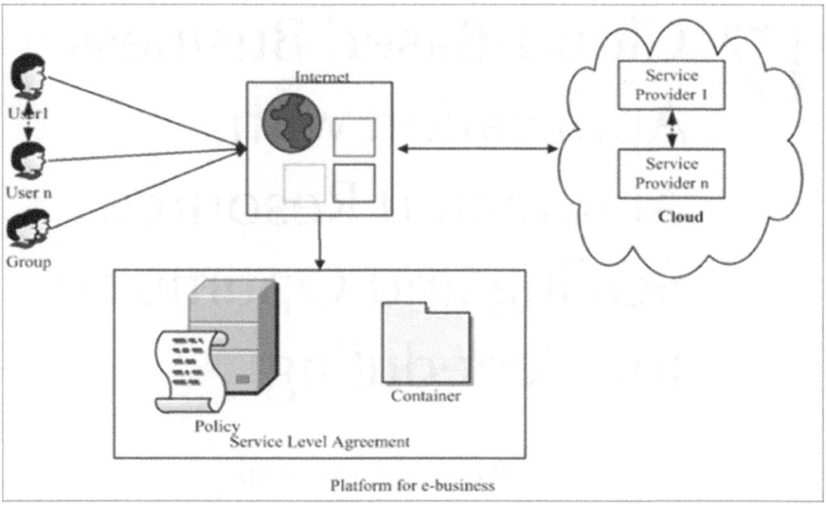

FIGURE 17.1 Service Level Agreement.

Figure 17.1 represents the Service Level Agreement (SLA) between the end user and Cloud Service Providers (CSPs). Either a single user or a group of users will establish a connection with the cloud to enhance their business through an online platform. The end users will sign an agreement between the user and the provider about the usage, processing speed, and so on. As a result, the scalability of computational resources is critical in meeting user expectations and service level agreement (SLA) requirements. SLA is a time and cost agreement between the user and the service provider [6]. CSPs support elastic and non-elastic resource provisioning methodologies to ensure SLA requirements [7, 8]. Elastic resource provisioning executes jobs by generating and deleting VMs at runtime for each job, resulting in excessive memory usage. Non-elastic resource provisioning, on the other hand, processes jobs using the VMs in the resource pool, resulting in resource underutilization and overutilization. Good scheduling algorithms schedule jobs to complete them on time and efficiently balance workloads across VMs. As a result, job and resource scheduling is complicated due to the need to balance the goals of CSPs, which seek to predict the economic consumption of computational resources and end users, who expect high-quality, cost-effective services [9, 10].

17.2 LITERATURE REVIEW

A scheduler is in charge of choosing jobs from a queue and allocating them to VMs so that they can be completed fairly and on time. Most schedulers use two algorithms, such as job scheduling and resource scheduling. Job and resource scheduling are critical for meeting the general needs of novice users as well as the goals of CSPs. To increase the number of jobs completed before the deadline, job scheduling algorithms change the order in which they are executed. After modifying the job execution order, resource scheduling assigns jobs to the best-suited VMs in the resource pool to meet the CSPs' objectives [11].

To meet the needs of inexperienced users, researchers have used various work scheduling algorithms based on various parameters such as deadline, time stamp, cost, makespan, and throughput. Initially, authors used the Earliest Deadline First (EDF) algorithm to process the jobs that have the earliest deadline [12]. This algorithm prioritizes order of jobs in such a way that an earlier deadline first to be completed, to increase the number of jobs completed before the deadline [12]. The EDF algorithm, however, does not preempt a job when an incoming job has an earlier deadline than the currently running job; thus, this algorithm has been expanded to the fully preemptive EDF algorithm (fp-EDF), which enables preemption of running jobs and processes jobs with the earliest deadline. Furthermore, the fp-EDF algorithm has been extended to the controlled preemptive EDF algorithm (cp-EDF) to prevent unwanted job preemption. Rather than superfluous job preemption, the cp-EDF algorithm conditionally preempts a running job based on constraints to increase resource utilization [13].

However, the fp-EDF and cp-EDF algorithms cause time starvation by scheduling jobs with the earliest deadlines in the queue. As a result, the Minimum Variation First (MVF) algorithm was developed to schedule jobs based on the minimum deviation, which is calculated by subtracting the deadline from the expected job execution time. The expected execution time of jobs can be calculated using their attributes, e.g., arrival time, length, and deadline [14]. In addition, the Adaptive Multilevel Scheduling System (AMSS) was proposed to schedule jobs based on the job parameters and the VM processing speed. Then it sequentially assigns jobs to the best-fitting VMs in the resource pool [15]. However, some jobs may not be complete before the deadline if this technique is used; thus, the Partial Critical Path (PCP) algorithm was proposed to process the jobs. The PCP algorithm is divided into two phases: distribution and planning. The distribution phase distributes the job deadlines to find the critical route, and the planning phase assigns a sub-deadline that maps the job to the least expensive service. Thus, optimal resource scaling is critical in elastic resource provisioning scenarios in cloud environments [16].

17.3 CHARACTERISTICS OF SYSTEM MODEL

17.3.1 Job Model

Many users have moved to cloud computing to process jobs cost-effectively as the pay-per-usage model [17], and on-demand resource provisioning have advanced. Because the number of jobs submitted to cloud systems varies over time, the incoming job rate in the cloud is difficult to predict; therefore, incoming jobs are considered aperiodic. In addition, knowledge about the job (i.e., a job's attributes) is also unknown earlier in the cloud system; thus, jobs are also referred to as 'apriori'. Generally, incoming jobs are stored in a queue $\left(J_q\right)$ and processed as a batch job. Jobs contain a bag of jobs $J_q = \left\{j_1, j_2, \ldots, j_n\right\}$, where n is the number of incoming jobs at a given time. Here, jobs are considered independent, i.e., the execution of a job in J_q is not dependent on the execution of another job. During job submission, the user submits two preliminary job parameters of the job $j_i(ls,d)$, where ***ls*** is the job length and ***d*** is the job deadline. Here, the job requests a single VM. In addition, after being assigned to a VM, the job is processed using only the allocated VM; thus, jobs are considered to be non-preemptive [18].

17.3.2 System Model

Cloud computing offers service [19] to the end users with the support of several CSPs.

$$Cc = \{CSP_1, CSP_2, \ldots, CSP_m\} \tag{17.1}$$

where m is the total number of CSPs. This work highlights how a CSP can optimally provide assistance to the end users. Most CSPs are embedded with several VMs.

$$CSP_i = \{VM_1, VM_2, \ldots, VM_\alpha\} \tag{17.2}$$

where α is the total number of VMs in CSP_i, and those are independent of each other. VMs in a CSP have two mandatory attributes, i.e., processing speed and storage $(VM_i(ps,s))$, where ps is the full processing speed of the VM, and s is the storage of the VM. The VM processing capacity varies according to the CSP policies. The CSP policy supports both homogeneous and heterogeneous environments. In a homogeneous environment, the VMs of a CSP have equal processing speed. The mathematical model is shown below.

$$VM_i(ps) = VM_j(ps) = VM_k(ps) \tag{17.3}$$

However, in a heterogeneous environment, VMs in a CSP have different processing speeds and are depicted as below.

$$VM_i(ps) \neq VM_j(ps) \neq VM_k(ps) \tag{17.4}$$

As a result, forecasting the job arrival rate is a more difficult task; thus, job scheduling is an important and complex task in cloud computing. Following that, jobs are processed using the available VMs. In general, the cloud system provides two types of resource provisioning. In the first category, jobs may be processed using available VMs in the resource pool; however, in this scenario, VMs are not utilized efficiently when there are few jobs. In contrast, the available VMs in the resource pool cannot process many jobs. Here, jobs may require more processing speed than can be provided by the VMs in the resource pool, and these jobs will only be completed after the deadline. This will have a negative impact on system performance, QoS, and user satisfaction. In the second category, a job can be processed at runtime by creating a generic VM based on user requirements. After a job is completed, VMs must be destroyed. The creation and destruction of a VM at runtime for each job have an impact on resource utilization. A high-priority job may arrive at a specific VM in the cloud system [19] during job execution. At this moment, the running job may be preempted before execution to complete the incoming job by the deadline. On the other hand, preempting the running job has an impact on resource utilization. Therefore, VMs are deployed in a two-tier architecture, i.e., foreground and background VMs. Thus, a single VM can process two jobs concurrently. Existing algorithms require additional time to prioritize and select the best VM from the resource pool. Henceforth, a new algorithm is required to process jobs optimally by integrating the VM in the resource pool and dynamically scaling the VM. For that reason, this paper proposes the OJSRS algorithm for cloud computing. To minimize overall processing time, the proposed algorithm processes jobs simultaneously. As a result, customer satisfaction and cloud service provider profits are enhanced.

17.4 PROPOSED SYSTEM DESIGN

Processing jobs are assumed to demand computational resources [20] in a cloud environment. The cloud environment comprises several CSPs that offer services to various clients over the Internet using a web interface. Each CSP has several data centers with several VMs to process a job. Depending on data center policy, the processing speed of VMs in the CSP may belong to a homogeneous or heterogeneous environment.

Figure 17.2 shows the design of the proposed OJSRS algorithm. The clients submit several requests which are treated as jobs and stored in a job queue named Q_{bj} . The jobs in the job queue are processed using the OJSRS algorithm in the cloud. Initially, the jobs are scheduled using tree-based job scheduling, and information about the VMs is stored in the Virtual Machine Monitor (VMM) [21]. After scheduling, few jobs are unmapped with the resources since the VMs in CSPs are incapable of completing the jobs within its deadline. In this scenario, the jobs are processed using a dynamic resource pool. In the proposed model, a Virtual Machine Monitor (VMM) stores information about the VMs in two queues. First, information about existing VMs is stored Q_{f-VMM}, and information about a created VM is maintained in a dynamic resource pool Q_{s-VMM}.

The minimum required processing speed of the job is computed according to the job's attributes, i.e., job length, job deadline, and job arrival time. The minimum speed of processing needed is denoted S_{er} and expressed by Equation 17.5.

$$j_x\left(S_{er}\right)=\frac{j_x\left(ls\right)}{j_x\left(d\right)};\forall jinQ_{bj} \tag{17.5}$$

where 'x' represents any job in the job queue. Depending on the minimum arrival rate of jobs, a fixed number of VMs is created based on the data center's policies, which are stored in Q_{f-VMM}. Using the existing VMs in the CSP, the minimum available processing speed of the CSP is denoted $CSP_{\min}\left(ps\right)$ and computed using Equation 17.6

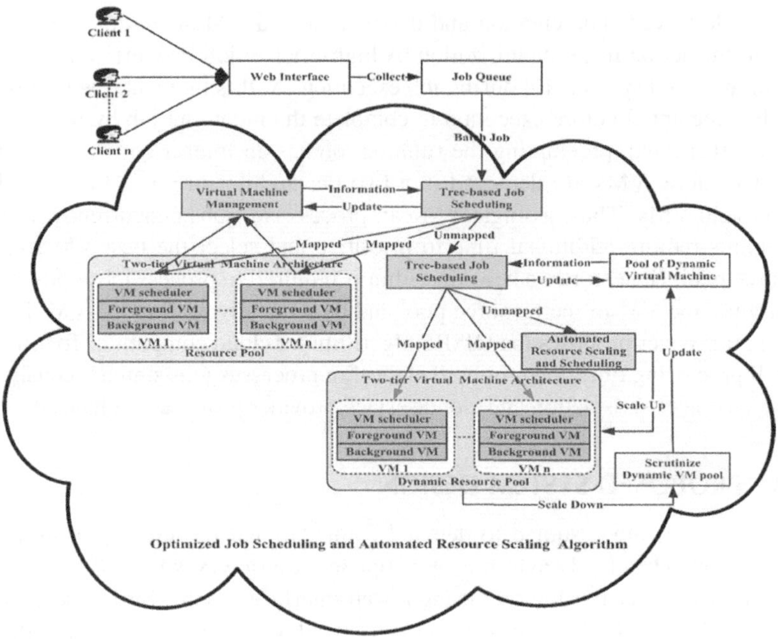

FIGURE 17.2 Proposed workflow of OJSRS algorithm

$$CSP_{\min}(ps) = \sum j_x(S_{er}) \;;\forall j\, in\, Q_{bj} \vee \sum VM_y(ps);\quad \forall VM\, in\, Q_{f-VMM} \tag{17.6}$$

The current computing capacity of the CSP is denoted $CSP_{current}(ps)$ and computed as given in Equation 17.7

$$CSP_{current}(ps) = \sum VM_y(ps);\ \forall VM\, in\, Q_{VMM} \tag{17.7}$$

Jobs are optimally mapped to VMs by creating a tree using the TJS technique [22]. However, some jobs cannot be processed using the VMs in after tree construction Q_{f-VMM}. Such jobs can be processed using the VMs in Q_{s-VMM}. Here, the TJS technique cannot allocate a VM to some jobs; therefore, a new VM is created at runtime using the AS technique, and the VM information is stored in Q_{s-VMM}. The TJS is described in detail in the following section.

17.4.1 Tree-based Job Scheduler

The TJS receives a batch of jobs from Q_{bq} that vary relative to time. Jobs in Q_{bq} are processed in a FCFS manner [23]. The idle processing speed of the VM is denoted S_{idle}. Initially, the TJS technique maps a job to the best suitable VM by comparing S_{idle} to S_{er}. It constructs a right-skewed binary tree $(RSBT)$ using $VM(S_{idle})$ in the Q_{f-VMM} of the VMM. Similarly, the TJS technique creates another right-skewed binary tree $(RSBT')$ using $VM(S_{idle})$ in Q_{s-VMM}. Here, the VMs in *RSBT* and *RSBT'* are considered parents in the TJS technique. After creation *RSBT* and *RSBT'*, the jobs are mapped to the respective VMs by comparing $j(S_r)$ to $VM(S_{idle})$, as given in Equation 17.8 Here, jobs are mapped to the best VMs that can complete the jobs by the deadline. The mapping of jobs to VMs is denoted $M(j_x, VM_y)$ and expressed as follows:

$$M(j_x, VM_y) = \begin{cases} 1; if j_x(S_{er}) \leq VM_y(S_{idle}) \\ 0; otherwise \end{cases} \quad \forall j in Q_{bq} \ \& \ \forall VM \tag{17.8}$$

where j_x and VM_y denote the corresponding job and VM, respectively. After mapping, $VM_y(S_{idle})$ is updated based on $j_x(S_{er})$. $VM_y(S_{idle})$ is computed as shown in Equation 17.9.

$$VM_y(S_{idle}) = \begin{cases} VM_y(S_{idle}) - j_x(S_{er}); & if M(j_x, VM_y) \\ VM_y(S_{idle}); & otherwise \end{cases} \tag{17.9}$$

After computing $VM_y(S_{idle})$, the job is mapped to the best suitable VM among all VMs in the resource pool using the TJS algorithm. Here, the scheduled time of the jobs is denoted t_s and stored before tree construction. The TJS algorithm [24] describes tree construction using $j_x(S_{er})$ and $VM_y(S_{idle})$.

Algorithm: TJS Algorithm
Input: Dataset $Q_{bq} = \{j_1, j_2, \ldots j_n\}, Q_{f-VMM}, Q_{s-VMM}$

Output: j_x *mapped with* VM_y

1. **Initially** $RSBT \leftarrow \varnothing, Q_{bq} \leftarrow \varnothing$
2. **Construct** $RSBT \ \forall VM \ in \ Q_{f-VMM}$ **// Root has minimum 'ps'**
3. **Built** $RSBT' \ \forall VM \ in \ Q_{s-VMM}$
4. **For each** $j \in Q_{bq}$ **do**
5. **For** $\forall VM \ in \ Q_{f-VMM}$ **do**
6. $M(j_x, VM_y) = \begin{cases} 1; & if \ j_x(S_{er}) \leq VM_y(S_{idle}) \\ 0; & otherwise \end{cases} \quad \forall j \ in \ Q_{bq} \ \& \ \forall VM$
7. $if \left(M(j_x, VM_y)\right) then$
8. $VM_y \leftarrow j_x$
9. Update $VM_y(S_{idle}) = \begin{cases} VM_y(S_{idle}) - j_x(S_{er}); & if \ M(j_x, VM_y) \\ VM_y(S_{idle}); & otherwise \end{cases}$
10. Else

11. **For** $\forall VM\ in Q_{S-VMM}$ **do**

12. $M(j_x, VM_y) = \begin{cases} 1; & if\ j_x(S_{er}) \le VM_y(S_{idle}) \\ 0; & otherwise \end{cases} \quad \forall j\ in Q_{bq}\ \&\ \forall VM$

13. $if\left(M(j_x, VM_y)\right) then$

14. $VM_y \leftarrow j_x$

15. Update $VM_y(S_{idle}) = \begin{cases} VM_y(S_{idle}) - j_x(S_{er}); & if\ M(j_x, VM_y) \\ VM_y(S_{idle}); & otherwise \end{cases}$

16. **End if**
17. **End for**
18. **End if**
19. End for
20. End for
21. End

The above TJS algorithm describes how the jobs are mapped to VMs. The mapping of jobs with the VMs is clearly indicated with the help of a skewed tree. The diagram is given in Figure 17.3.

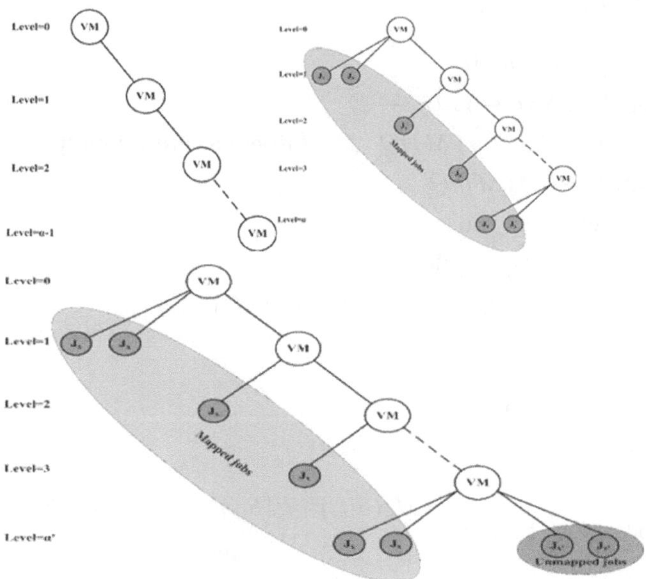

FIGURE 17.3 Tree construction using the TJS technique.

Figure 17.3 shows the construction of a tree using the TJS technique and elaborates how jobs are created and mapped to the VMs. Here, jobs are mapped to the VMs based on the following criteria. The VMs in $RSBT$ are considered parents, and jobs are inserted as either a right or left a child by comparing S_{er} and $VM(ps)$. As a result, a new job is inserted to the left sub-tree if S_{er} is less than or equal to $VM(ps)$; otherwise, it is compared to the processing speed of the VM in the next level of the tree. Here, $RSBT$ is created with several VMs (α) in the resource pool, and the $RSBT$ tree has $\alpha - 1$ levels. Occasionally, they are not mapped to any VMs in $RSBT$, and these VMs are passed to $RSBT'$. Thus, each level of the tree has only one parent and one or more leaves. After the tree is created using the TJS technique, jobs are assigned to the corresponding VMs. Here, each VM has a queue (Q_{VM}) to store the corresponding jobs. The VM at each level processes all corresponding left child jobs; however, VMs in the α-level of the $RSBT$ tree cannot be processed. Thus, these jobs may be processed using VMs in the dynamic resource pool (Q_{s-VMM}). In addition, VMs at the level $\alpha' - 1$ of the $RSBT'$ tree cannot process all jobs in the right child of the VM, where α' is the total number of VMs in Q_{s-VMM}. Thus, VMs must scale up to process the right child of the node at the $\alpha' - 1$ level using auto-scaling.

17.4.2 Automated Resource Scaling and Scheduling

Due to the unpredictable arrival of jobs, VMs must scale up and down dynamically to minimize the overall expenditure of the CSP. Jobs assigned to the right side of the node at level $\alpha' - 1$ cannot be processed using VMs in Q_{f-VMM} and Q_{s-VMM}. These unmapped jobs are moved to a queue Q_{nj}. Here, the CSP must scale up the VMs to process and complete the jobs by the deadline. The processing speed of a scaled VM depends on the data center's policies in a homogeneous environment. However, in a heterogeneous environment, VMs are created based on the minimum required processing speed of the job. The minimal additional processing speed of unmapped jobs in Q_{nj} is denoted A_{uj}, which is manipulated using various job attributes of the jobs and computed as in Equation 17.10.

$$A_{uj}(ps) = \sum j(S_{er}); \qquad \forall j \, in \, Q_{nj} \tag{17.10}$$

The total processing speed of scaled-up VMs is denoted as A_{uj}. The number of VMs created in a homogeneous environment depends on the data center's policies $\left(i.e. \eta = \frac{A_{uj}(ps)}{VM(ps)} \right)$. However, in a heterogeneous environment, the number of VMs made depends on S_{er} of the jobs. After completing VMs, information about the VMs is updated in Q_{s-VMM}, and the jobs are processed in the VMs. Occasionally, a sparse number of jobs may arrive in the cloud system. In such cases, the VMs in the resource pool must be scaled down to avoid wasting computational resources. Here, the job arrival rate is denoted $J_{arrival\,rate}$ and computed in Equation 17.11.

$$J_{arrival\,rate} = \sum_{i=1}^{n} j_i(S_{er}) \tag{17.11}$$

The minimum required processing speed of the CSP $\left(i.e.Q_{f-VMM}\right)$ is scaled down depending on the job arrival rate and calculated in Equation 17.12.

$$CSP_{\min}\left(ps\right)=\begin{cases}J_{arrival\,rate};if\left(J_{arrival\,rate}<CSP_{\min}\left(ps\right)\right)\&\,VM\,is\,idle\\CSP_{\min}\left(ps\right); \qquad otherwise\end{cases} \tag{17.12}$$

17.4.3 Performance Analysis

Initially, the number of created VMs is based on the job arrival rate, which is always less than the number of VMs in a homogeneous environment [25] because a single VM can process and complete multiple jobs before the deadline. However, the number of VMs in a heterogeneous environment is initially equal to the number of incoming jobs. Here, the number of VMs is smaller than the number of incoming jobs. In the proposed OJSRS model, jobs are processed in an FCFS manner. The proposed algorithm simultaneously maps the job to the best-fit VM using the job's required processing speed and the idle processing speed of the VM by constructing the tree. Incoming jobs are sometimes sparse and dense; thus, the VMs in the data center must be scaled up and down dynamically to improve resource utilization.

17.4.4 Job Delay Time

As mentioned previously, the job arrival rate varies over time in cloud computing [26]; therefore, jobs were generated randomly without following any conventional scale. The proposed OJSRS model focuses on minimizing the job wait time to increase user satisfaction. Existing scheduling algorithms [27], e.g., FCFS, EDF, and AMS, sequentially select the best-fit VM to process jobs using the VMs in the resource pool. However, the proposed OJSRS model constructs a tree and concurrently maps a job to the best-fit VM in either Q_{f-VMM} or Q_{s-VMM} by comparing the '*ps*' and S_{er} of the job to minimize scheduling time. Occasionally, some jobs are not allocated to any VM when it $J_{arrival\,rate}$ is greater than the processing rate of the VMs in the resource pool. As a result, such jobs can only complete execution after the deadline due to either an inadequate number of VMs in the resource pool or the inability of the VMs in the resource pool.

Figure 17.4 represents the job delay time in a homogeneous environment. The proposed OJSRS algorithm shows that the job delay time is less than the other algorithms because the number of VMs is increased based on the arrival rate of the jobs. The FCFS, EDF, and AMS process the jobs only with the help of available VMs. Since the OJSRS integrates the auto-scaling of resources based on the arrival rate of the jobs, the job delay time is approximately 50 % less than the other algorithms by scaling up VMs based on the data center's policy. The job delay time in heterogeneous environments [28]. With the proposed model, the VMs are scaled up based on the minimum required job processing speed $J_{arrival\,rate} \geq \sum VM\left(ps\right)$ in VMM to

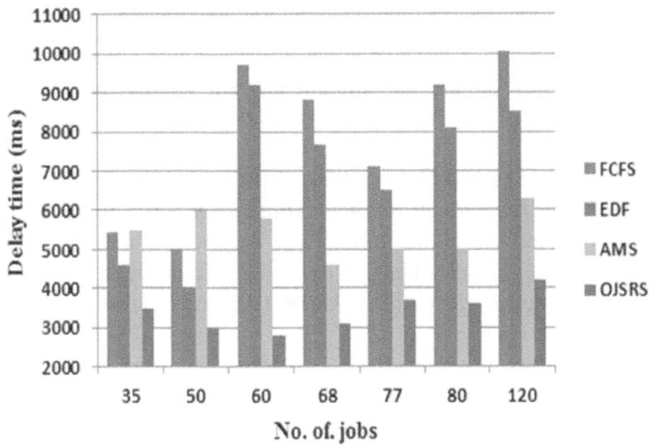

FIGURE 17.4 Job delay time inhomogeneous environment.

minimize the job wait time. The VMs are created for the unmapped jobs in the right-skewed tree.

17.5 CONCLUSION AND FUTURE WORK

To complete incoming jobs before the deadline, cloud systems face various challenges. An algorithm is proposed in this research work to minimize SLA violations, improve resource utilization, mitigate job wait times, increase user satisfaction, and improve CSP performance. The proposed OJSRS algorithm schedules jobs based on the job arrival rate algorithm and constructs a tree that concurrently bounds jobs to the corresponding VMs in the resource pool. Some jobs cannot be mapped to any VMs in the resource pool after binding. Such jobs can be processed by dynamically scaling up the VMs in the CSP based on either data center policy or the minimum required job processing speed. Furthermore, VMs are deployed as foreground and background VMs in a two-tier architecture. One or more jobs are bound to the corresponding VM; thus, jobs are prioritized based on different job attributes to avoid deadline violations. Following this prioritization process, an incoming job may have a higher priority than currently running jobs. In such situations, the incoming job is processed in the background VM while the running job is conditionally halted. The proposed OJSRS algorithm outperforms existing scheduling algorithms by mitigating job deadline violations. It converges job scheduling with resource scalability to improve resource utilization by 5 percent and the performance of the cloud system but not in real-time environment. It is planned to improve the proposed algorithm to process the dependent jobs more efficiently. In the future the job scheduling task can be made optimal by considering different environments with respect to job availability, resource utilization, and job completion delay using different advanced deep learning models over Fog and Edge computing.

REFERENCES

1. G. T. Reddy, K. Sudheer, K. Rajesh and K. Lakshmanna, "Employing data mining on highly secured private clouds for implementing a security-as-a-service framework," *Journal of Theoretical and Applied Information Technology*, vol. 59, no. 2, pp. 317–326, 1714.
2. K. Dinesh, G. Poornima and K. Kiruthika, "Efficient resources allocation for different jobs," *International Journal of Computer Applications*, vol. 56, no. 10, pp. 30–35, 1712.
3. R. M. Swarna Priya, S. Bhattacharya, P. K. R. Maddikunta, S. R. K. Somayaji, K. Lakshmanna, et al., "Load balancing of energy cloud using wind driven and firefly algorithms in internet of everything," *Journal of Parallel and Distributed Computing*, vol. 142, pp. 16–26, 1717.
4. C. Li, A. Raghunathan and N. K. Jha, "A trusted virtual machine in an untrusted management environment," *IEEE Transactions on Services Computing*, vol. 5, no. 4, pp. 472–483, 1712.
5. S. Chaisiri, Bu-Sung Lee and D. Niyato, "Robust cloud resource provisioning for cloud computing environments," *IEEE International Conference on Service-Oriented Computing and Applications (SOCA)*, pp. 1–8, 1710.
6. A. Stanik, M. Koerner and O.Kao, "Service-level agreement aggregation for quality of service-aware federated cloud networking," *IET Networks*, vol. 4, no. 5, pp. 264–269, 1715.
7. J. Moura and D. Hutchison, "Review and analysis of networking challenges in cloud computing," *Journal of Network and Computer Applications*, vol. 60, pp. 113–129, 1716.
8. I.F. Senturk, P. Balakrishnan, A. Abu-Doleh, K. Kaya, Q. Malluhi, et al., "A resource provisioning framemwork for bioinformatics applications in multi-cloud environments," *Future Generation Computer Systems*, vol. 78, pp. 379–391, 1718.
9. B. Palanisamy, A. Singh, L. Liu and B. Langston, "Cura: A cost-optimized model for mapreduce in a cloud," *IEEE 27th International Symposium on Parallel & Distributed Processing (IPDPS)*, pp. 1275–1286, 1713.
10. B. Palanisamy, A. Singh and L. Liu, "Cost-effective resource provisioning for mapreduce in a cloud," *IEEE Transactions on Parallel and Distributed Systems*, vol. 26, no. 5, pp. 1265–1279, 1715.
11. S. Vakilinia, M.M. Ali and D. Qiu, "Modeling of the resource allocation in cloud computing centers," *Computer Networks*, vol. 91, pp. 453–470, 1715.
12. M. Sathish Kumar and M. Iyapparaja, "Improving quality-of-service in fog computing through efficient resource allocation," *Computational Intelligence*, vol. 36, no. 4, pp. 1–21, 1717.
13. R. Van den Bossche, K.Vanmechelen and J. Broeckhove, "Online cost-efficient scheduling of deadline-constrained workloads on hybrid clouds," *Future Generation Computer Systems*, vol. 29, no. 4, pp. 973–985, 1713.
14. D. Komarasamy and V. Muthuswamy, "Job scheduling using Minimum Variation first algorithm in cloud computing," *IEEE Sixth International Conference Advanced Computing (ICoAC)*, pp. 195–198, 1714.
15. D. Komarasamy and V. Muthuswamy, "Deadline constrained adaptive multilevel scheduling system in cloud environment," *KSII Transactions on Internet and Information Systems (TIIS)*, vol. 9, no. 4, pp. 1302–1317, 1715.
16. S. Abrishami, M. Naghibzadeh and D.H. Epema, "Deadline-constrained workflow scheduling algorithms for infrastructure as a service clouds," *Future Generation Computer Systems*, vol. 29, no. 1, pp. 158–169, 1713.

17. R. N. Calheiros and R. Buyya, "Meeting deadlines of scientific workflows in public clouds with tasks replication," *IEEE Transactions on Parallel and Distributed Systems*, vol. 25, no. 7, pp. 1787–1796, 1714.
18. D. Huang, C. Zhu, H. Zhang and X. Liu, "Resource intensity aware job scheduling in a distributed cloud," *China Communications*, vol. 11, no. 14, pp. 175–184, 1714.
19. M. Iyapparaja and Bhanupriya Sharma, "Augmenting SCA project management and automation Framework," *IOP Conference Series: Materials Science and Engineering*, vol. 263, pp. 1–8, 1717.
20. P. Yi, H. Ding and B. Ramamurthy, "Cost-optimized joint resource allocation in grids/clouds with multilayer optical network architecture," *Journal of Optical Communications and Networking*, vol. 6, no. 10, pp. 911–924, 1714.
21. L. Zhou and W. Haohong, "Toward blind scheduling in mobile media cloud: Fairness, simplicity, and asymptotic optimality," *IEEE Transactions on Multimedia*, vol. 15, no. 4, pp. 735–746, 1713.
22. L. Zhou, Z. Yang, J.J. Rodrigues and M. Guizani, "Exploring blind online scheduling for mobile cloud multimedia services," *IEEE Wireless Communications*, vol. 17, no. 3, pp. 54–61, 1713.
23. Y. H. Lee, S. Leu and R. S. Chang, "Improving job scheduling algorithms in a grid environment," *Future Generation Computer Systems*, vol. 27, no. 8, pp. 991–998, 1711.
24. F. López-Pires, B. Barán, L. Benítez, S. Zalimben and A. Amarilla, "Virtual machine placement for elastic infrastructures in overbooked cloud computing datacenters under uncertainty," *Future Generation Computer Systems*, vol. 79, pp. 830–848, 1718.
25. S. Deva Arul and M. Iyapparaja, "Social internet of things using big data analytics and security aspects-a review," *Electronic Government, an International Journal*, vol. 16, no. (1–2), pp. 137–154, 1717.
26. T. Chen and B Rami, "Toward a smarter cloud: self-aware autoscaling of cloud configurations and resources," *Computer*, vol. 48, no. 9, pp. 93–96, 1715.
27. M. Iyapparaja, "Effective feature selection using hybrid GA-EHO for classifying big data siot," *International Journal of Web Portals*, vol. 12, no. 1, pp. 12–25, 1717.
28. Z. Xiao, Q. Chen and H. Luo, "Automatic scaling of internet applications for cloud computing services," *IEEE Transactions on Computers*, vol. 63, no. 5, pp. 1111–1123, 1714.

18 Addressing Data Security and Privacy Challenges

A Critical Analysis of Convolutional Neural Network (CNN) Applications

Mudassir Khan, Rupali Atul Mahajan, Rajesh Dey, and Shaik Karimullah

18.1 INTRODUCTION

With the advancement of technology, applications of convolutional neural networks (CNNs) are rapidly growing in a variety of fields including education, healthcare, finance, autonomous vehicles, and security systems. These applications are often fed with sensitive data such as personal identification documents, bank statements, medical images, etc. This information could be misused if falls into the wrong hands. Hence, the security and privacy of data used in CNN systems are of utmost importance[1]. CNNs have shown remarkable success in a variety of problems ranging from image and video recognition to natural language processing. Market players have been investing heavily in CNN-based commercial products. For example, Facebook uses CNNs to automatically tag people in photos uploaded by users. Google and Apple use CNNs for on-device face matching to provide user authentication based on facial images. Following the success of CNNs, users tend to rely on CNN-based services to conduct their desired tasks, resulting in an increasing transfer of sensitive data to service providers. Despite the great advances and successful applications of CNNs, there exist vulnerabilities that can be exploited to harm the interests of users or service providers [2]This study investigates the data security and privacy challenges arising in the applications of CNN and proposes the corresponding solutions. First, a comprehensive overview is provided of the possible threats to data security and privacy in CNN applications. Then, methodologies are critically reviewed that can safeguard data against the identified vulnerabilities. Finally, the future research directions are discussed. By providing a comprehensive overview of a contemporary topic, this study will be of great interest to researchers in the fields of machine learning and data security & privacy. With the rapid progress of Artificial

DOI: 10.1201/9781003567585-18

Neural Networks (ANN), Deep Neural Networks (DNN) with many hidden layers have been applied to various fields for data classification.

18.1.1 Background and Significance

Continuous advances in information and communication technologies have driven the rapid growth of human society into a new era of ubiquitous computing, networking, and digital sensing. A large volume of data is being generated, collected, stored, and processed in a wide range of applications, including social networks, smart cities, finance, healthcare, and environmental monitoring. Among them, sensitive data that involves personal privacy and confidential information is either being shared or considered for sharing by many organizations. On the one hand, shared data can create new business opportunities and improvements in services and operational efficiencies. On the other hand, it can also bring a number of risks and challenges associated with data security and privacy [3]. Data security refers to the protective measures against unwanted and unauthorized access and data tampering. Data privacy is the collection and sharing control of personal and sensitive data. With the increasing digitization, networking, and proliferation of data, the likelihood, scale, and impact of data breaches have continuously increased, thus making data security and privacy one of the biggest challenges in the digital age. Convolutional neural networks (CNNs) have emerged as a fundamental machine learning technology, powering a wide range of applications, products, and services in both private and public sectors. Recently, there have been growing concerns about the vulnerability of CNNs to malicious attacks that can intentionally induce CNNs to make wrong predictions. These attacks can be categorized into two types: evasion attacks that target the inference process of CNNs, and poisoning attacks that target the training process of CNNs [4]. With the increasing adoption of CNNs for processing sensitive data, such as personal information and financial records, CNN data security and privacy issues have emerged as an existential threat. Protection of data security and privacy is especially critical when CNNs are implemented on open and untrusted platforms, such as cloud servers, edge servers, or shared servers.

18.1.2 Research Objectives and Scope

The proliferation and integration of information and communication technologies (ICTs) have brought significant transformations in the development, delivery, and access of information services. The convergence of various ICTs, such as the internet, telecommunications, broadcasting, satellite, and computer networks, has led to the birth of new media, characterized as digital, smart, automated, networked, and web-based [5, 6]. The advancement of ICTs has also paved the way for the growth of big data, artificial intelligence, the internet of things, and cloud computing, offering novel opportunities and prospects for economic growth and social development. Along with these advancements, concerns over the security and privacy of data have emerged as critical issues worldwide. The global massive data breaches and privacy infringements have triggered intensive discussions on data security and privacy at different levels, including national, organizational, and individual levels. At the national level, several regulations, acts, and policies have been proposed and enacted to ensure the security and privacy of data. At the organizational level, various security frameworks, protocols, tools, and technologies have been developed and deployed to secure data and information systems. At the individual level, different precautions and protection measures have been adopted to minimize the exposure of personal data. Nevertheless, high-profile data breaches continue to occur. Understanding the potential vulnerabilities in the data security and privacy of the currently emerging technologies and applications has become an essential task with critical significance. Convolutional neural networks (CNNs) are widely applied and extensively studied deep learning models. CNNs have been successfully utilized in various fields and industries, such as automotive, healthcare, finance, and information. Although considerable efforts have been devoted to investigating CNN applications, concerns over data security and privacy have received limited attention. Addressing this gap is crucial for the responsible and sustainable development and deployment of CNN applications. Against this backdrop, a critical examination of the data security and privacy issues and challenges of CNN applications is conducted. First, the existing state-of-the-art CNN models and frameworks are reviewed, and the internationally emerging CNN applications are identified. Then, the potential threats, vulnerabilities, and challenges concerning data security and privacy are analyzed in detail. Finally, the future research directions are proposed to inspire and guide future inquiries. The intended contribution of this study is threefold: (1) it identifies the emerging CNN applications and analyzes their usage scenarios and types of data; (2) it systematically examines the possible threats, vulnerabilities, and challenges concerning data security and privacy; and (3) it proposes critical future research directions to stimulate further investigation. By systematically addressing these concerns, this study contributes to the ongoing efforts to ensure the responsible and sustainable development and deployment of CNN applications in society.

18.2 DATA SECURITY AND PRIVACY FUNDAMENTALS

Data security and privacy are very important issues and particularly pressing where large quantities of sensitive data are involved. As a result, those concerns are most palpable in the context of social network data, healthcare data, and in the context of financial institutions. Additionally, since the pace of technological development is now so rapid, such concerns have brought forth new challenges, particularly because most tools required to achieve the aims of data privacy and data security can barely keep up with the progress in technology. Not least, the great strides being made with deep learning neural network computations continually set new benchmarks. That is why the issues raised in this work are so critically important and logically require a sharp focus on the trade-offs that exist with machine understanding business intelligence [7].

The availability of various data analytics techniques, together with the advanced computational power of big data processing systems, often enables us to better understand data and make better data-driven decisions. However, the use of mining results in the extraction of information that can be exploited by criminals, or perhaps even those who would use it to misinform others with nefarious intent. Unfortunately, because the data involved is confidential and sensitive, its misuse could result in significant harm—ranging from identity theft and financial fraud to the deliberate spread of misinformation by malicious actors.

18.2.1 Key Concepts in Data Security and Privacy

A critical requirement for all communication protocols consists of ensuring that the information exchanged is kept confidential. This is achieved by using secure communication protocols. Although security concepts are an ancient notion, they have special relevance in modern communication infrastructures, characterized by high-speed digital devices exchanging information wirelessly. Data privacy, concepts, and enforcing protocols are more recent ideas that apply security practices concerning the possession, use, and disclosure of information by individuals. The idea that humans are entitled to privacy, either physical or informational, is well established. Using electronic communication, either wired or wireless, either to transfer data or to provide services, raises a number of related issues about data protection. Data concerns determine that information exchanged using communication protocols must be kept confidential.

Despite the enormous advances in technology during the last decades, man's brain remains slightly more powerful than any machine ever built. However, computers can now store and analyze large amounts of data quicker than the information available in the human brain. In many scenarios, problems that would be impossible or impractical to be treated by humans are dealt with by computers.

18.2.2 Importance of Data Security and Privacy in CNN Applications

In general, the field of big data for data-driven scientific discovery is experiencing a rapid utility increase. This fact comes from the new generation of deep learning. In this sense, it is reported that CNN is a machine learning technique that is implemented today in various big data and analytics systems. These technologies have the ability to help in searching for information in different ways. Deep learning systems are trained with a series of algorithms inherent in the CNN to observe and make decisions based on previous experience. However, the most critical security and privacy concerns arise from the centralized model for deep learning techniques, which include data vulnerabilities and the need to access an extremely high level of sensitive data. First, the approach allows sensitive data to be generalized; any centralized repository can expose clients to privacy theft, creating a marketplace to develop benign models to obtain data. Second, considering that the majority of the training datasets used by large companies contain sensitive personal information and are created from public and private sources, access to the data exposes non-security issues associated with privacy, which can lead to a wide spectrum of data from various sources being used illegitimately [8].

18.3 CONVOLUTIONAL NEURAL NETWORKS (CNNS): AN OVERVIEW

Artificial neural networks (ANNs) are computational models inspired by biological neural networks. They consist of interconnected artificial neurons that process and transmit information. ANNs learn from data by adjusting connection weights, allowing them to approximate complex functions and solve various tasks. A specific class of feedforward networks is called convolutional neural networks (CNNs). CNNs are a type of artificial neural network designed to process structured grid data, such as images. Images represent 2D grids of pixels, and CNNs exploit their spatial structure and correlation to automate feature extraction. CNNs have gained significant popularity in recent years, particularly in the field of computer vision. In 2012, a breakthrough CNN architecture significantly outperformed other computer vision techniques in the ImageNet Large Scale Visual Recognition Challenge (ILSVRC). Since then, CNNs have become the dominant approach in computer vision, leading to state-of-the-art performance in various tasks. CNNs have also gained attention outside of computer vision, being used in natural language processing and even in non-image data applications like genomics and drug design. CNN revolutionized image recognition, transforming how computers interpret visual data. CNN has a unique architecture that mimics the human visual cortex's hierarchical structure. It comprises three main types of layers: convolutional, pooling, and fully connected layers. Each layer type has its characteristics and functions, performing specific operations on input data [9]. The architecture uses convolutional layers to process visual data, and pooling layers reduce data dimensionality while retaining essential features as shown in Figure 18.1.

The structure of CNN varies based on the task at hand, but it generally consists of a sequence of pairs of convolutional and pooling layers followed by fully connected layers. Imagine a natural scene captured by a camera. The scene is represented as an array of pixel values, with each pixel consisting of red, green, and blue values. This representation forms a three-dimensional array (height, width, depth) that serves as the input to the CNN architecture. Convolutional layers implement convolution operations, which are mathematical operations that merge two sets of information. These layers learn a set of filters that are convolved with the input data during the training process. Each filter detects specific features in the input, such as edges, shapes, or textures. The output of the convolution operation is called the feature map, representing the detected features at different locations in the input. Multiple filters are learned in each convolutional layer, resulting in multiple feature maps.

To ensure appropriate dimensions, parameters such as stride and padding need to be defined. The feature maps produced by the convolutional layers are then passed through the pooling layers, which perform down-sampling operations. These layers reduce the dimensionality of the feature maps while retaining the most critical information, making the representation invariant to small translations of the input. Pooling layers help reduce overfitting, computational complexity, and memory

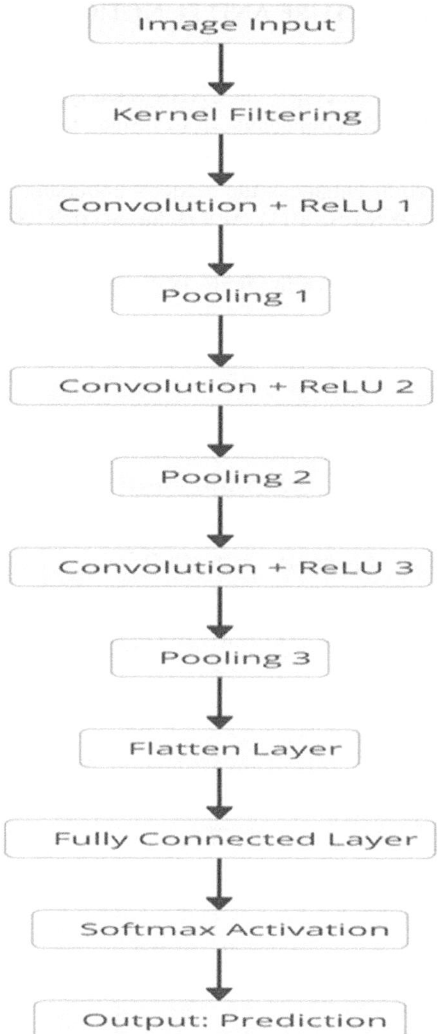

FIGURE 18.1 An overview of convolutional neural network (CNN) flow diagram.

requirements. The most common operation in pooling layers is max-pooling. Finally, fully connected layers are added at the end of the architecture. Each neuron in this layer is connected to all the activations in the previous layer, similar to traditional multilayer perceptrons. Each fully connected layer employs a SoftMax activation function to produce class scores, determining the probabilities of the input belonging to different classes. CNNs are widely used for image recognition and processing tasks, transforming how businesses operate [10].

18.4 BASIC ARCHITECTURE AND FUNCTIONALITY

Focusing on the basic architecture and functionality, this subsection delves deeper into the individual layers that constitute CNNs. Each type of layer—convolutional, activation, pooling, and fully connected—is examined to elucidate its specific functions in the network. By detailing the process by which CNNs extract features from input images, this portion enhances the reader's understanding of how these networks operate (O'Shea & Nash, 2015). As a technique that continues to gain traction in data security and privacy applications, it is crucial to develop a comprehensive understanding of CNNs as shown in Figure 18.2.

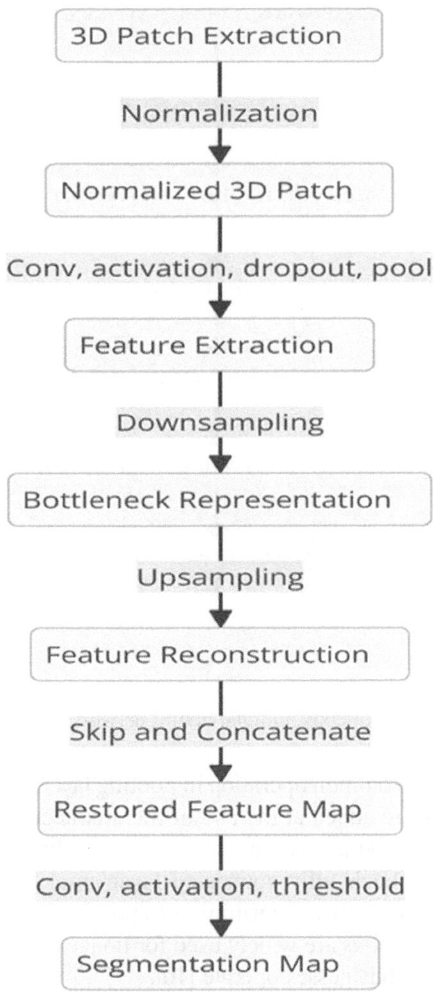

FIGURE 18.2 3D medical image segmentation pipeline flow chart.

Moreover, CNNs are sophisticated technologies that, while not exceedingly new, deserve the thorough attention given here. The 2012 ImageNet Large Scale Visual Recognition Challenge marked a watershed moment for CNNs[11]. CNNs triumphed by a remarkable margin, outperforming more traditional image classification methods, thus prompting a significant shift in the research community's focus toward deep learning architectures. In the years following this success, CNNs became the dominant architecture for a wide array of computer vision applications. With such widespread adoption and continued research, efforts are directed toward novel applications for CNNs.

18.5 APPLICATIONS IN IMAGE RECOGNITION AND PROCESSING

Convolutional neural networks (CNNs) have emerged as one of the most impactful technologies in recent years, finding applications across diverse sectors. This section focuses on image recognition and processing, highlighting a range of CNN applications, from facial recognition in smartphones to medical image analysis. Each example illustrates how reliance on CNN technology is growing in various fields, thus presenting practical use cases with transformative impacts. CNNs automate image recognition and processing tasks, significantly improving efficiency and accuracy. However, challenges persist, such as variability in image quality, compromised effectiveness in low-quality images, and the need for large training datasets. Each application example critically discusses how these challenges are addressed as shown in Figure 18.3. Given the sensitivity of the data involved, security and privacy considerations are also highlighted. As CNN applications continue to expand, it is vital to understand these dimensions to mitigate risks. Thus, this section explores the

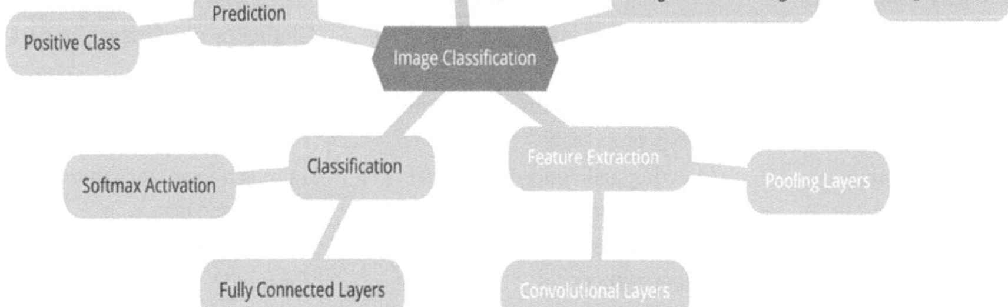

FIGURE 18.3 CNN-based brain tumor classification pipeline.

diverse applications of CNNs in image recognition and processing, showcasing their transformative impacts .

CNNs are increasingly used in everyday applications, transforming how images are analyzed and interpreted. For instance, CNNs can accurately identify and tag users in photos, allowing the social media platform to offer "photo-tagging" functionality. CNNs are also widely used in smartphones for locking mechanisms based on face detection. The technology detects and recognizes a user's face using the phone's front camera, enabling unlocking. Similarly, many online shopping sites use CNN-based object detection techniques that locate a specific object in a scene. Medical imaging relies heavily on CNN applications, such as detecting tumors in MRI scans [12].

18.6 DATA SECURITY CHALLENGES IN CNN APPLICATIONS

In contemporary context, CNN applications address various needs across multiple domains with high accuracy and reliability. However, CNN applications expose several data security challenges that need to be considered. Primarily, adversarial attacks are discussed as significant data security challenges against CNN applications. Adversarial attacks are manipulative inputs generated by malicious entities aiming to alter the normal output of trained CNN models [13]. Recent research revealed that CNNs are highly vulnerable to adversarial attack inputs, even with a slight alteration of sample inputs. Based on the knowledge of adversaries, adversarial attacks are classified as white-box, grey-box, and black-box attacks. On the contrary, considering the ignorance of adversaries, threat models are categorized into L_p norm-bounded, data-dependent, and unnoticeable perturbations. Reviewing the attack vectors of adversarial attacks against CNN models highlights the pressing need to develop effective defense mechanisms as shown in Table 18.1.

In addition to adversarial inputs, the risk of privacy leakage of data passing through CNN applications is identified as another data security challenge against

TABLE 18.1
Comparison of Data Security Challenges in CNN Applications

Security Challenge	Description	Example Scenario
Adversarial Attacks	Manipulating input data to mislead CNN predictions	Altering medical images to trick diagnostic AI
Privacy Leakage Risks	Exposure of sensitive data through CNN processing	Revealing patient data in healthcare AI models
Model Poisoning	Injecting malicious data to alter model behavior	Training a facial recognition AI with biased data
Data Tampering	Unauthorized modifications to CNN input/output	Altering financial transaction data for fraud
Regulatory Compliance Issues	Challenges in adhering to data privacy laws	GDPR compliance in AI-driven surveillance

CNN applications (Mireshghallah et al., 2020). Privacy leakage primarily risks the data used for training and inference processes in CNN applications. It is significant as it may threaten individual privacy and organizational integrity, particularly for CNN applications in healthcare and banking sectors. In CNN modeling, member attributes of each data instance are protected or are privately kept with an outside service provider. In such circumstances, the risk of privacy leakage is critically high against the forwarded data of CNN modeling services, data tampering attacks, deniable proof, and model exchange attacks. Reviewing several emerging privacy leakage attacks against CNN applications highlights various possible means to compromise privacy leakages and the urgency of developing and implementing defense mechanisms. Overall, CNN applications significantly impact modern and future computational usage.

18.7 ADVERSARIAL ATTACKS AND DEFENSES

This subsection will focus on adversarial attacks against CNN applications. Attacks against CNNs are crafted to compromise the integrity of the underlying data. These attacks typically involve the creation of adversarial examples, which are data inputs that have been modified in a way that is undetectable to a human but misleads a CNN. The added changes are so minor that a person would still recognize the input correctly, yet they are enough to make the CNN misclassify it. For example, a picture of a panda may be altered slightly to become misclassified as a gibbon. Adversarial examples are at the core of a new class of attack vectors against machine learning: evasion attacks. In evasion attacks, the attacker exploits a previously trained model by providing it with adversarial examples. Evasion attacks against CNNs typically target image-based CNNs that are used for applications such as object detection, image recognition, or scene classification. On the opposite side, there are attacks that compromise the training process of a machine learning model. In these poisoning attacks, the attacker manipulates the model training, for instance, by injecting poisoned examples into the training dataset. Researchers showed that CNNs used in malware detection can be attacked by both types of adversarial attacks. Defense strategies against adversarial attacks are grouped into two categories: (1) strategies that have been proposed in the machine learning literature to defend against adversarial examples and (2) innovations specifically developed to safeguard CNNs. Among the first group of techniques, adversarial training, model regularization, and input preprocessing are the most notable ones. In adversarial training, the model is trained on both clean and adversarial examples, improving robustness against the specific type of attack. Model regularization involves adding a penalty on the model parameters or decision function to reduce sensitivity to perturbations, such as L2-regularization or adding noise to the input. For the second category of techniques, the focus is on innovations specifically developed for CNNs. In general, the proposed defense strategies need to be tailored to the specific architecture of the neural network. There are also generic defense strategies that apply to all deep neural networks, such as using ensemble methods. These data security challenges are not limited to the lab environment but have already been encountered in the wild with real-world incidents. For example, the security of biometric systems has been compromised with the presentation of fake samples. A thorough understanding and the development of effective defense strategies for adversarial perturbations are still open research problems [14].

18.8 PRIVACY LEAKAGE RISKS

Privacy is a critical issue in deep learning applications involving user data, such as computer vision and natural language processing tasks handled by convolutional neural networks (CNNs). Despite obvious privacy leakage risks, organizations using CNN applications are often unaware of the challenges and responsibilities of preserving user data privacy. Privacy leakage refers to the possible exposure of sensitive information from trained CNNs, data transmitted for inference, and data used during processing. Privacy leakage risks can occur at any stage of data handled by CNNs, encompassing the training dataset, model parameters, and data processed for inference [15]. For example, the rapid development of deep learning techniques in healthcare raises significant concerns about the confidentiality of patient data used in research [16]. In CNN applications involving user data, compliance with privacy protection laws poses a challenge regarding how data are used by CNNs. In particular, questions arise concerning users' rights and whether their data have been appropriately handled or utilized, shedding light on ethical concerns in CNN applications (Miriyala Reddy & Miriyala, 2020) as shown in Table 18.2.

These challenges and concerns need to be unpacked to bring attention to the difficulties organizations face in ensuring data privacy and confidentiality when

TABLE18.2
Privacy-Preserving Techniques for CNN Security

Technique	Description	Strengths	Limitations
Homomorphic encryption	Enables computation on encrypted data without decryption	Strong security, preserves confidentiality	High computational cost
Differential privacy	Adds noise to data to prevent individual identification	Effective against privacy breaches	Reduces data utility
Secure multi-party computation	Enables collaborative computation without revealing private data	Suitable for distributed AI systems	Requires significant computational resources
Federated learning	Trains models across multiple devices without centralizing data	Enhances privacy by keeping data local	Vulnerable to poisoning attacks
Blockchain for data integrity	Uses decentralized ledgers to ensure data authenticity	Prevents unauthorized modifications	High storage and processing demands

applying CNNs. CNN applications can often be essential to an organization's services or assistance [17]; however, limitations arise in ensuring the privacy of user data fed into CNNs outside the organization. Several practical examples are provided to illustrate the severity of privacy leakage risks in CNN applications, often resulting in the unintended exposure of sensitive information. Organizations using CNNs to process sensitive data need to be aware of these risks and consider effective measures to prevent potential privacy breaches. It is vital to think through and develop innovative solutions to prevent possible privacy breaches before continuing the widespread implementation of CNNs, as the challenges of ensuring data privacy and confidentiality continue to grow [18].

18.9 PRIVACY-PRESERVING TECHNIQUES IN CNN

Addressing the growing concern of confidentiality breaches, privacy-preserving techniques that allow processing sensitive data without exposing it are drawing increasing attention. There are two popular approaches: data encryption and data perturbation. Homomorphic encryption, a special encryption technique preserving mathematical properties, falls under the former category. Encrypted data can still be computed by specific functions without revealing the underlying data. This means that sensitive data can be kept secret all the time during the whole analytic process. In CNN-based applications, input data and network weights should be kept confidential [18]. Data perturbation methods add noise to the original data to hide sensitive information from attackers. It has been shown that, as long as individuals' data contributions in a dataset are not disclosed, any analysis on this dataset will guarantee that individuals cannot be tracked or re-identified, even when the output of the analysis is public. This concept is called differential privacy [19]. Input data should be protected in CNN training processes, while either input data or model parameters should be protected in CNN inference processes. Privacy-preserving designs based on homomorphic encryption or differential privacy have been proposed to resolve such confidentiality challenges. Quantization-related confidentiality challenges and potential solutions are discussed first. Then two representative scenarios are examined, where either input data or model parameters should be kept secret: one CNN model encrypted with the Paillier homomorphic cryptosystem outperforms the other using the ReLU activation function, and for CNN models encrypted with the same homomorphic cryptosystem, garbled circuits achieve better performance than HEAAN.

18.10 HOMOMORPHIC ENCRYPTION

Homomorphic encryption is a privacy-preserving technique that can be applied to CNNs. The underlying principle of homomorphic encryption is that computations are performed on encrypted data directly, without the need for decryption. Once the data is encrypted using a homomorphic encryption scheme, the user can send the encrypted data to the recipient. The idea of homomorphic encryption was proposed by Rivest, Adleman, and Dertouzos in 1978 [12]. In 2009, Gentry proposed the first fully homomorphic encryption scheme, using lattice-based cryptography. Since then, practical fully homomorphic encryption schemes have been developed, such as BGV, FV, and YASHE. One of the earliest works to support neural networks with homomorphic encryption was proposed by Gentry, Sahai, and Waters in 2011. They designed a novel encryption scheme that allowed for classifying encrypted data using a neural network. A practical approach to performing inference in neural networks with homomorphic encryption was proposed later by Chollet in 2015. This work implemented a simple feed-forward neural network with the Paillier encryption scheme. Paillier homomorphic encryption supports only addition operations on encrypted data, making it impossible to implement CNNs that require convolutions (Choi et al., 2024). A framework for building CNN inference services with encrypted inputs based on the PyTorch library was presented. This work, built on the SEAL library, allowed users to deploy 2D CNNs by converting input data into a two-dimensional matrix, where convolution operations were replaced by pointwise matrix multiplications. However, this method limits the variety of deployable CNN models since all layers should be compatible with an encrypted data structure. To address this issue, a homomorphic encryption-based CNN architecture is proposed, which implements the CNN architecture itself in a homomorphic encryption scheme. This architecture eliminates the need for ensuring compatibility with a diverse range of CNN models using homomorphic encryption. Experimentation shows that the proposed architecture outperforms the leading 2D CNN inference architecture, PyCrCNN, in terms of inference time. Additionally, it outperforms the architecture in processing concurrent images, which is an essential feature for high-demand cloud applications.

18.11 DIFFERENTIAL PRIVACY

Differential privacy is a highly regarded strategy that safeguards individual privacy when using sensitive data in applications powered by convolutional neural networks (CNNs). Simply put, it guarantees that the outcome of a mechanism cannot be reliably linked back to an individual's input, even with access to auxiliary information about that individual. Mathematically, a randomized mechanism M satisfies ε-differential privacy if, for all datasets D and D′ differing on at most one element, where D′ can either be obtained from D by removing an element or vice versa, and for all $S \subseteq R$, the range of M:

$$\Pr[M(D) \in S] \leq \exp(\varepsilon)\Pr[M(D') \in S].$$

Thus, ε is a privacy budget that controls the strength of differential privacy protection. Lower ε values indicate greater privacy preservation (Demelius et al., 2023). The opportune choice of ε values depends on the specific use case and the dataset's severity level. In general, $\varepsilon < 1$ is considered strong privacy, while $\varepsilon > 3$ indicates that privacy is mostly preserved. This mechanism can be enforced either at the dataset or individual data levels. The dataset-level mechanism involves incorporating noise into the dataset before exposing it to analysis. This noise is usually Gaussian or Laplace, dependent on the sensitivity (i.e., maximum variation on output) of the function $f : D \rightarrow R$ used to derive data utility. In contrast, individual data-level differential privacy primarily focuses on adding noise directly to the individual's data being exposed and is commonly denoted as local differential privacy. Still, it is paramount to note that adding noise can distort the dataset, rendering it less useful.

18.12 ETHICAL CONSIDERATIONS IN DATA SECURITY AND PRIVACY

As artificial intelligence (AI) technologies continue to evolve, transforming various sectors, it becomes imperative to consider their ethical dimensions concerning data security and privacy. This need arises because the application of AI, including convolutional neural networks (CNNs), often involves handling extensive digital data. Thus, the onus is on researchers and developers to establish ethical guidelines governing the use of such data within the designed technologies (Radanliev et al., 2024). The algorithmic decisions made by CNNs can significantly influence marginalized groups, especially when determining an individual's eligibility for essential services. It is therefore vital to interrogate the data on which CNN models are trained, as this can exacerbate existing biases. Furthermore, transparency and accountability are crucial aspects to consider when developing and deploying CNNs. Questions arise regarding who is responsible for the algorithmic outcomes of CNN technologies and whether stakeholders, such as organizations, have a duty to disclose the data used in model development.

18.13 BIAS AND FAIRNESS IN CNN MODELS

This subsection delves into the critical issues of bias and fairness concerning the use of CNN models. With a focus on computer vision applications, it assesses the potential for biases to unintentionally permeate CNN training processes, resulting in outcomes that may discriminate against certain individuals or groups. Attention is first drawn to the model's training processes, which encompass considerations like dataset composition, model architecture, and post-processing methods. Each stage is analyzed to identify the underlying sources of bias that contribute to fairness issues. Then, via a more technical examination of the CNN output, the implications of bias on societal perceptions and decision-making processes are scrutinized. Ultimately, the urgency of addressing these concerns is highlighted by summarizing compelling real-world examples of biased CNN outputs [1].

18.14 TRANSPARENCY AND ACCOUNTABILITY

Transparency and accountability are critical themes with respect to the applications of CNNs. Users generally have a desire for as much transparency as possible in AI systems, requesting clear and accessible documentation indicating how a given model operates and makes decisions. This is especially true when users' data are at stake. The level of trust in technology significantly increases when users have access to basic information regarding how it works (Tanzib Hosain et al., 2023). As such, organizations are called upon to provide transparency regarding their AI products to foster a sense of trust among users. In regard to CNNs, transparency is particularly imperative at the level of model development, including data usage and potential biases found within it. Without transparency in these areas, accountability cannot be enforced. Without accountability, there can be no deterrence against the possible malicious or unethical use of data; rather, opaque algorithms would only exacerbate this issue, compounding preexisting trust problems. The idea is that, in addition to being intrinsically interpretable by users, systems should be designed such that outside audits of their decision-making processes are possible. Regarding accountability, the responsibilities of various stakeholders are outlined, including developers, organizations, and end-users. Ultimately, regulations and other frameworks are viewed as essential for fostering accountability in AI development.

18.15 CONCLUSION

As the technical landscape evolves, so do threats to confidentiality and privacy. One of the most pressing threats is the ongoing proliferation and sophistication of attack vectors aimed at sensitive collected data, precipitated by the rise of the Internet of Everything (IoE) and pervasiveness of artificial intelligence (AI). Data privacy breaches can translate into severe liability for organizations in terms of litigation, regulatory fines, competitive disadvantage, and reputational harm. CNN technologies can both exacerbate privacy harm and serve as catalysts for defensive remediation strategies. Though CNNs have recently been implicated in data security and privacy vulnerabilities, there is currently very little academic examination of these issues. This critical literature review will begin to fill that gap by considering how CNN technologies simultaneously threaten and safeguard data security and privacy (Mireshghallah et al., 2020). As AI technology proliferates across sectors, effort must be spent characterizing interrogative mechanisms and prescriptive policy frameworks appropriate to practitioners in novel CNN contexts (Radanliev et al., 2024). Karimullah et al. (2024) developed an Integrated Method for Detecting Lung Cancer via CT Scanning via Optimization, Deep Learning, and IoT Data Transmission. Kaur et al. (2024) find the high-accuracy lung disease classification via logistic regression and advanced feature extraction techniques

REFERENCES

1. Cherepanova, V., Nanda, V., Goldblum, M., Dickerson, J. P., & Goldstein, T. (2021). Technical challenges for training fair neural networks. *arXiv*. https://arxiv.org/abs/2102.06764
2. Choi, H., Kim, J., Kim, S., Park, S., Park, J., Choi, W., & Kim, H. (2024). UniHENN: designing more versatile homomorphic encryption-based CNNs without im2col.
3. Demelius, L., Kern, R., & Trügler, A. (2023). Recent advances of differential privacy in centralized deep learning: a systematic survey.

4. Hussain, A. Karimullah, S. Khan, M., & Shaik, F. (2025). Comprehensive methodology to the detection and classification of emotion in human face using EMOTE-Net. *Fusion: Practice and Applications,* 19(1), 10–22. https://doi.org/10.54186/FPA.190102.
5. Jain, N., Nandakumar, K., Ratha, N., Pankanti, S., & Kumar, U. (2018). Efficient CNN building blocks for encrypted data.
6. Karimullah S, Khan, M, Shaik F, Alabduallah B and Almjally A (2024) An integrated method for detecting lung cancer via CT scanning via optimization, deep learning, and IoT data transmission. *Frontiers in Oncology,* 14, 1435041. https://doi.org/10.3389/fonc.2024.1435041
7. Kiran, A., Elseed Ahmed, A., Khan, M., Babu, J., & Kumar, B. (2025). An efficient method for privacy protection in big data analytics using oppositional fruit fly algorithm. *Indonesian Journal of Electrical Engineering and Computer Science*, 37(1), 670–679. https://doi.org/10.11591/ijeecs.v37.i1.
8. Kumar K., K. & N V S L Narasimham, D. (2023). Enhancing radiology diagnosis through convolutional neural networks for computer vision in healthcare.
9. Mireshghallah, F., Taram, M., Vepakomma, P., Singh, A., Raskar, R., & Esmaeilzadeh, H. (2020). Privacy in deep learning: a survey.
10. Miriyala Reddy, S. & Miriyala, S. (2020). Security and privacy preserving deep learning.
11. Nanda, V., Dooley, S., Singla, S., Feizi, S., & P. Dickerson, J. (2020). Fairness through robustness: investigating robustness disparity in deep learning.
12. Oseni, A., Moustafa, N., Janicke, H., Liu, P., Tari, Z., & Vasilakos, A. (2018). Security and privacy for artificial intelligence: opportunities and challenges.
13. O'Shea, K. & Nash, R. (2015). An introduction to convolutional neural networks.
14. Radanliev, P., Santos, O., Brandon-Jones, A., & Joinson, A. (2024). Ethics and responsible AI deployment.
15. Kaur, S., Gupta, S., Gupta, D., Juneja, S., Nauman, A., Khan, M., Husain, I., Islam, A., & Mallik, S. (2025). High-accuracy lung disease classification via logistic regression and advanced feature extraction techniques. *Egyptian Informatics Journal,* 29, 100596. https://doi.org/10.1016/j.eij.2024.100596
16. Tang, K., Zeng, C., Fu, Y., & Zhu, G. (2022). Security analysis of social network topic mining using big data and optimized deep convolutional neural network.
17. Tanzib Hosain, M., Hasan Anik, M., Rafi, S., Tabassum, R., Insia, K., & Mehrab Siddiky, M. (2023). Path to gain functional transparency in artificial intelligence with meaningful explainability.
18. Weerasena, H. & Mishra, P. (2023). Revealing CNN architectures via side-channel analysis in dataflow-based inference accelerators.
19. Yuan, M. J., Zou, Z., & Gao, W. (2024). Bi-CryptoNets: leveraging different-level privacy for encrypted inference.

19 Cybernetic Shield and IoT-Based Vibration Fencing for Wildlife Protection

Arun V., Kalaiselvi P., Gopikadevi J.A., Lochini J., Aishwarya T., and Yoga Lakshmi S.

19.1 INTRODUCTION

The expansion of agricultural lands into wildlife habitats has raised human–wildlife conflicts, and consequently, there has been more loss of crops, livestock depredation, and ecological disturbances. With human populations still on the rise and moving into virgin territories, wildlife is pushed into agricultural farms, where farmers incur [1] losses and introduce more danger to animal populations. The increasing rate of urbanization and agricultural expansion in the forest region not only threatens biodiversity but also decreases natural habitats. This leads to pushing wildlife into human habitation and agricultural farms. These conflicts demand for an urgent implementation of sustainable wildlife management policies that minimize risks [2–4] while promoting peaceful coexistence between people and wildlife. Unless checked, the conflicts would worsen, resulting in irretrievable loss of biodiversity and long-term environmental costs.

People have used different methods to stop wild animals from entering farms, such as building electrical fences, using strong-smelling chemicals, or even harming the animals. But these ways can cause problems. Electrical fences [5–7] cost a lot of money and even result in animal death. Chemicals can harm animals, plants, and the soil. It is not right to hurt animals because it can disturb nature. Old methods are not always safe or ethical. We need new ideas that can protect farms and animals at the same time. These new ways should also be good for the environment. Following the same traditional methods can be dangerous for animals and farming in the future. So, we should use better and ethical methods that help both humans and animals.

In response to these demands, we suggest a cybernetic IoT-based vibration fencing system, providing automated wildlife deterrents, real-time monitoring, and green operation. Using controlled vibrations as a deterrent, the system provides an environmentally friendly, non-lethal alternative to existing practices [8–10]. Through the

 DOI: 10.1201/9781003567585-19

application of IoT technology, the system can adapt to the dynamic nature of environmental conditions and maximize energy consumption. Scalability of this system guarantees applicability to a wide range of agricultural landscapes, rendering it a desirable move toward the integration of agricultural security and ecological conservation. Integration of smart fencing technologies not just improves efficiency but encourages the adaptation of sustainable and humane agricultural practices, giving rise to a new paradigm for modern agricultural protection.

19.2 LITERATURE SURVEY

Human–wildlife conflict is considered to be a serious issue, especially when it comes to agricultural farms, where wildlife intrusion leads to massive crop damage, economic damage, and alteration in local ecosystems. Electric fences have been used as deterrents for centuries; however, they come with a huge number of disadvantages, including high maintenance, high power requirements, and ethical issues related to their impact on animal welfare [3]. The research has shown that deterrent systems, such as electric fencing, rarely yield enduring results and often end up with unforeseen ecological consequences, such as habitat fragmentation and changed migration routes [5]. Therefore, researchers focused their attention on the development of advanced technologies that offer ethical, humane, and harmless solutions for reducing human–wildlife conflicts with minimal negative impacts on wildlife.

The eleAlert system has been proposed as a new method of managing human-elephant conflict (HEC), especially in areas such as Sri Lanka, where elephant intrusion typically leads to large-scale destruction of agricultural crops [1]. Dissimilar to conventional electric fences that operate on static barriers, the eleAlert system employs Remote Transmitting Units (RTUs), bridge spaces, and Location Identification Tags (LITs) to identify and monitor breaches in electric fencing. One of the principal strengths of the system is the capability to deliver real-time alerts through automated data transmission of the animal movement, eventually reducing the necessity of manual [11] checks and allowing faster response times by farmers [12] and wildlife management authorities. The system also initiates geospatial analytics and predictive modeling, which enable risk assessment and predictive warnings where conflicts are likely to happen. Initial results have shown that the eleAlert system gradually decreases the cases of human-elephant conflict, which will confirm its potential as a valuable, scalable, and cost-effective solution for the intrusion of wildlife into agricultural farms.

Recent developments in machine learning and AI-based [13] detection models have further optimized real-time wildlife monitoring systems. Studies have shown that CNNs, especially VGG-16-based models, attain up to 77% correct species identification of major wildlife species such as elephants, wild boars, and buffalo [6]. AI-based models process large volumes of image and sensor data, separating species to reduce false alarms and enhance intervention methods. Transfer learning methods and LoRa-based communication systems have also optimized response time and detection accuracy, allowing farmers to receive real-time notifications through wireless networks [6]. Adaptive algorithms allow AI-based detection [14] models to adjust automatic deterrence interventions through learning and adaptation to species-specific activity patterns over time, minimizing occurrences of ineffective deterrence and maximizing overall efficiency of smart fencing solutions.

The use of vibration sensors in the detection of elephants has attracted a lot of attention as a harmless and effective method in reducing human-elephant conflict. Researchers [12] showed the effectiveness of vibration-based detection systems in identifying patterns of elephant movement and preventing damage to crops. The detection systems based on vibrations use ground sensors that detect vibrations caused by approaching elephants along the fence line and send an early warning to farmers and wildlife authorities. Vibration sensors are not similar to conventional electric fences since they utilize a more humane approach without directly making contact with animals. The research shows the advantages of the technology as being low maintenance, versatile across landscapes, and providing real-time data of the movement of the animal. The use of IOT and wireless communication improves accuracy and response time in these detection systems that are used to prevent wildlife deterrence. This shows that vibration-based detection has the ability to play an essential part in modern cybernetic shielding and wildlife conservation measures, resulting in both ethical and sustainable solutions in minimizing human–wildlife conflict [2].

The wildlife deterrence systems based on IOT are made to execute advanced tracking, geospatial mapping, and mobile alert applications, which enhance surveillance and intervention measures. GPS tracking, ArcView GIS software, and cloud monitoring systems provide an overview of animal movement near the crops, which allows researchers and farmers to predict possible intrusion risks by the wildlife and take preventive measures against them [7]. Mobile applications are developed using React Native and Firebase, providing real-time alerting to users, allowing them to take rapid action when animals are found approaching guarded agricultural plots. Cloud computing also facilitates bulk processing of data, which optimizes decision-making, allows pattern identification, and facilitates the formulation of long-term mitigation strategies [8]. Predictive analytics, driven by AI-based models, can analyze historical movement data to forecast future wildlife patterns, thereby allowing optimization of the location and functionality of deterrence systems.

A sustainable method to prevent animals from entering the farm is by using solar-based fencing [2]. These fences work with solar panels that capture the energy from the sun, which is then stored in batteries for continuous operation during nighttime or in times when there is a lack of sunlight. These types of fences are used especially in areas where the absence of continuous access to electricity occurs. Charge controllers are used to manage the power safely and securely. To scare animals without harming them, these fences consist of flashing lights, and they also use noises that sound like bees. Some systems also have PIR sensors to detect movement and GSM modules to send alerts. It uses solar energy, which is better than the electrical fences, which are not easily damaged by large animals. Overall, this solar fencing is a better way to reduce conflicts between humans and animals.

19.3 METHODOLOGY

The vibration fencing system is aimed in a way that it offers an innovative, ethical, and harmless solution to keep wild or domestic animals away from agricultural farms without harming them. Instead of using electric shocks, this system uses vibrations that make the animals feel uncomfortable, so they go away. It uses an ultrasonic sensor, a vibration motor, and IoT for real-time monitoring. The ultrasonic sensor can able to sense the animal when it comes close, less than 30 cm. These ultrasonic sensors emit high-frequency sound waves and calculate the time for the echo to return after reflection off an object. When the sensor detects the movement of a distance less than 30 cm, the system will trigger the vibration motor to produce the vibration (Figure 19.1).

When an animal is detected, the vibration motor is turned on, which produces vibration to scare the animals from the farm. The LoRaSX1278 module is used for long-range communication and helps send messages over long distances, so the farmers can get alerts in real time. The system also has NodeMCU controllers that act like mini-computers, and they help the fence work ethically and smartly without hurting the animals.

This system works by using dual power sources: solar power and electricity. This helps the system keep working at all times, even without much sunlight. The NodeMCU controller reads the data from the sensor, and when it detects an animal, it turns on the vibration motor. These vibrations scare off the animal, actually without harming it. The LoRa SX1278 module ensures long-range communication between the fence and the control system, enabling real-time alerts and remote control. The vibration motor gets powered by Arduino, which turns it ON and OFF when needed. To measure the distance, the system uses TRIG and ECHO pins. When an animal comes in contact, the sensor detects it using the formula, and the motor is turned on to vibrate. The system is designed to reduce unnecessary activations. This helps the system to save energy and prevents animals from getting used to vibrations over

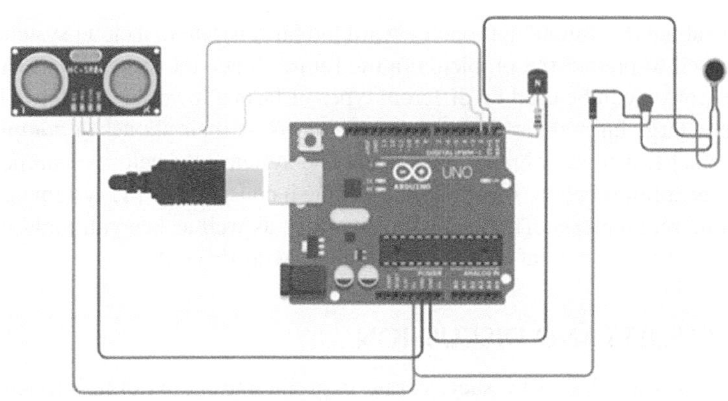

FIGURE 19.1 Prototype of vibration fencing.

time. The sensor collects data about the animal's movements along the fence when the vibration is detected. This information is sent to the cloud system, where it can be used for research to improve the system's performance. To make it more effective, we can include some additional features such as flashing LEDs, notification alerts, real-time messages, and sound alarms. The system can also use smart sensors that can learn and adjust to different weather or environmental conditions. By combining these sensors with the vibration system, the solution becomes more ethical, resulting in reduced conflicts between humans and wildlife.

19.4 SYSTEM ARCHITECTURE

The Vibration Fencing System is an advanced way to stop wild or domestic animals from entering the agricultural farms, and it successfully does that without causing any harm to the animal. The sensing layer has ultrasonic sensors placed along the fence that keep scanning for animals. The three main layers for the system is the sensing layer, the control layer, and the actuation layer. These sensors send out high-frequency sound waves, and by checking how long it takes for the waves to bounce back, they can figure out how far the animal is. This information then goes to the control layer. In the control layer, there are NodeMCU microcontrollers that study the signals to decide if it's time to activate the fence. If an animal is close enough, the actuation part gets triggered. That's when vibration motors and things such as flashing LED lights and sound alarms turn on to scare the animal away. The system also uses machine learning to tell the difference between normal movement and an actual threat. The system works fast and smart because of IoT technology, which lets it send messages to monitoring stations far away. It uses LoRa SX1278 modules to quickly send alerts when animals are nearby. The system mostly runs on solar panels, which get energy from the sun. But if there isn't enough sunlight, it has a backup battery that turns on.

It is also energy efficient; it works based on the weather and how often the animals enter the fence. The vibration motor in the system produces vibration along the fence, which is used to scare off animals from the farm without hurting them. All this data about the animal approach toward the farm is sent to a cloud system, which can be used to predict the problems in the future. Since the system is modular and easy to grow, it can be used on different types of farms. It works well for both small farms and large farm areas. Farmers can get alerts on their phones or computers, so they can act fast if an animal comes near. The system also helps researchers learn about how animals behave near farms. By using it consistently, the system gets better at spotting real dangers. This helps protect crops as well as keep animals safe. The cybernetic shield is a great way to solve human–wildlife conflict.

19.5 RESULTS AND DISCUSSION

The system showed an 80% success rate in predicting the wildlife intrusion in the crops. It represents the efficiency of the system in crop protection without harming any animals. The use of ultrasonic sensors and vibration motors successfully repelled animals and provided an eco-friendly and effective alternative to electrical fencing. Farmers also observed a drastic change in crop loss and confirmed the real-time usage. The trials also showed that the system succeeded in a wide range of environmental conditions. Even in heavy rains and overpowering winds, the system showed great results. Significantly, the dual-power source showed a smooth operation even when sunlight was low. The data collected from the agricultural farm resulted in a 40% reduction in human–wildlife conflict.

The system latency testing portrayed an effective response rate and accuracy in the movement of the animals. It also succeeded in detecting accurate real-time movement when any wildlife approached the fence. We tested the vibration fencing against the electrical fencing, which highlighted the need for an ethical solution. This electrical fencing causes physical harm and injuries to the animal and, on rare occasions, even death. The requirement for a cost-effective and ethical way to protect the crops was high. We decided to work on the demand for solutions for human–wildlife conflicts. The future advancement in vibration fencing would be a notification or alert system.

19.6 LIMITATIONS

Although the system works well, there are a few limitations that need to be fixed before implementing it in other applications. One of those relies on solar power, which may not work well in areas with less sunlight. To overcome this, we should implement it in such a way that it can support a mix of energy sources that make sure the system always works. Even though the system is mostly accurate, sometimes it gives false alarms or misses something, so better AI and sensor adjustments are needed. Also, the vibration-based alarms don't work the same for all animals, so the system should be able to adjust to different animals over time. Environmental factors, such as thick plants or bad weather, can affect how well the sensors work; careful sensor placement and protection need to be implemented.

Another limitation is the initial cost of installation, which may deter small-scale agricultural producers even when long-term operation and maintenance cost savings can be achieved. Simple finance options or economically more suitable designs can enhance affordability. Inadequate connectivity in rural areas can also reduce the timely receipt of alerts, and offline data storage and other forms of communication are essential. Although the system is efficient in restricting the conflicts between humans and wildlife, the adaptability of the system in different ecosystems needs to be researched further. These limitations would have to be addressed through technology and policy interventions so that cybernetic shielding and IoT-based vibration fencing can scale up and be effective in wildlife conservation.

19.7 CONCLUSION

This chapter describes the Cybernetic Shield and IoT-Based Vibration Fencing System, an ethical solution to preventing human–-wildlife conflict and guaranteeing sustainable agriculture. The system uses ultrasonic sensors, vibration detection mechanisms, and Internet of Things connectivity, providing a harmless and energy conservation-based means of protecting agricultural assets. Combining the capabilities of long-distance communication and real-time automation enhances accuracy in intruder detection and deterrence activation, thus minimizing false alarms and maximizing overall system performance. When compared to traditional electric fencing, this technology highly improves animal welfare, lowers operating costs, and promotes environmentally friendly farm practices. The flexibility of the fence system in conforming to the diverse terrains and the remote monitoring capability make the system scalable and feasible for mass deployment in wildlife-prone areas.

REFERENCES

1. Wijesinghe, L., Siriwardena, P., Dahanayake, S., Kasthuriratne, D., Corea, R. and Dias, D., 2011, October. Electric fence intrusion alert system (eleAlert). In *2011 IEEE Global Humanitarian Technology Conference* (pp. 46–50). IEEE.
2. Using the vibration sensors to detect elephants-2018. May 2018. *Kasem Bundit Engineering Journal,* 8 (Special Issue).
3. Whiting, T.L., 2016. Pain in human and non-human animals caused by electricity. *The Canadian Veterinary Journal*, *57*(8), p. 883.
4. Human-Wildlife conflict in India: a review of economic implications of loss and preventive measures. 2016. *Indian Forester,* 142 (10), pp. 928–940.
5. Electrical injuries in animals: causes, pathogenesis, and morphological findings. 2016. 53(5), pp. 1018–1029.
6. Wild animal detection using a machine learning approach and alerting using LoRa communication. *2022 International Conference on Smart Generation Computing, Communication and Networking.*
7. Solar-based fencing and animal detection system for agricultural fields. February 2019. *JETIR.*
8. Real-time animal detection and prevention system for crop fields. June 2021.
9. Towards an IoT-based notification system for agriculture electric fence. *2020 IEEE Student Conference.*
10. IoT-based animal trespass identification and prevention system for smart agriculture.*2023 7th IEEE International Conference.*
11. The performance of electric fences as elephant barriers in Amboseli, Kenya. April 2008. *South African Journal of Wildlife Research 38*(1)
12. Liu, X., Yang, T. and Yan, B., 2015, November. Internet of Things for wildlife monitoring. In *2015 IEEE/CIC International Conference on Communications in China-Workshops (CIC/ICCC)* (pp. 62–66). IEEE.
13. Lakshmi, K.M., Raja, C., Sreekanth, D. and Renuka, N., 2020. Security for protecting agricultural crops from wild animals using GSM technology. *Journal of Shanghai Jiaotong University, 16*(7), pp. 1007–1172.
14. Meera, S., Priyadharshini, K., Varshitha, P.V. and SaiCharitha, R., 2022, November. Animal detection alert system. In *2022 1st International Conference on Computational Science and Technology (ICCST)* (pp. 538–542). IEEE.

20 Enhanced Object Detection in Underwater Imaging Using Deep Learning Method

Priyanka Gorai, Anuj Singh, and G. Balamurugan

20.1 INTRODUCTION

Underwater images suffer from poor visibility, low contrast, and noise, making object detection challenging. Traditional methods used for object detection struggle to detect objects accurately, so we have used a customized approach for better underwater object detection [1]. Our chapter aims to improve the underwater object detection through a two-step process that first enhances image quality and then detects objects. In the first stage we use CLAHE, which helps in solving visibility issues. In the next stage, we use the Faster R-CNN model, which identifies and classifies the objects in enhanced images. To improve the effectiveness of the model, we have used data augmentation, transfer learning, and tuning. To handle challenging examples, we have additionally used Online Hard Example. To achieve more accurate results, we have used GIOU loss that refines object boundaries [2]. For overlapping detections, Soft Non-Maximum Suppression (Soft-NMS) is implemented. Using image improvements and faster-RCNN models, we have created a more accurate and effective way to find objects underwater.

20.2 LITERATURE REVIEW

According to the researchers, they have researched focusing on image enhancement by applying various algorithms.The several techniques that have been produced by showing and appealing the output. Let's traverse some remarkable algorithms and methods from the preceding research papers. Wei Song et al. suggested that the algorithms and methods which have been used for the image enhancement and detection part join a different upgraded approach [1–4]. Their methodology points to two important areas, such as storing the data of the image and recognizing the color of the image in each part and each pixel. To obtain this, they have used an operation that is called "white balancing," which fits the color, and by that, they can be seen

DOI: 10.1201/9781003567585-20

as more genuine and also show the technique that depends on the statistical novel models that can enhance the underwater environment and quality of the image. Yue Zhang et al. [5] have identified the problem and started labeling the problem happening to the enhancement of the image quality, such as red tint and noise present in the dataset of the image. The enhanced process of the image ensures that the output images are not only cleared from the system but also show the detailed capture of the true colors and details of the underwater environment. Aashi Singh et al. [1] initiated influential underwater images effectively. Their approach mainly focused on the improvement of the quality and transparency of these images efficiently [7].

20.3 PROPOSED MODEL

The methodology begins with data collection, utilizing a well-verified dataset from Kaggle, which has been validated by researchers. The process involves several steps, as illustrated in Figure 20.1.

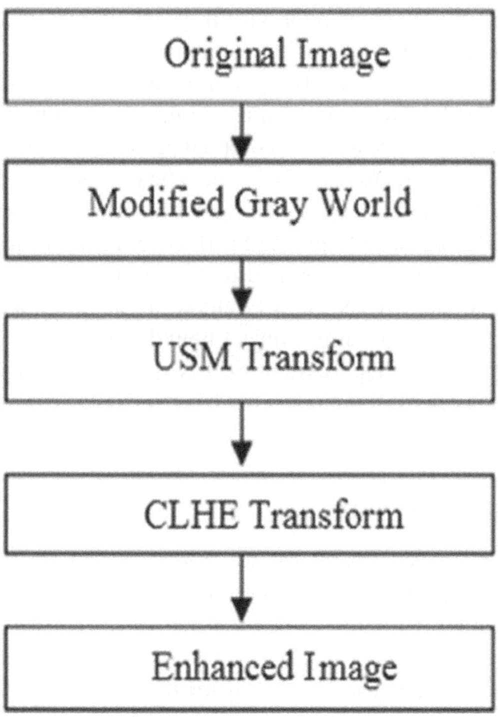

FIGURE 20.1 Steps for image enhancement.

20.4 ALGORITHMS USED FOR IMAGE ENHANCEMENT

20.4.1 White Balancing Using the Gray World Algorithm

White balancing using the Gray World Algorithm helps improve the colors in underwater images, ultimately making object detection easier. Underwater images often look too bluish or greenish as the water absorbs red light, which makes colors look unnatural and can hide important features. White balancing fixes the color and makes it look more like what we would see on land [6]. The gray world algorithm is one of the most commonly used algorithms that helps to adjust the colors so that none of the colors such as blue, red, or green stands out too much and makes it look like soft gray. To identify how much each color in the image is dominating we calculate the average color values for red, green, and blue channels across. Then we adjust each color channel in such a way that the image has too much blue; it will reduce the blue intensity [7], and if it has too little red, it will boost the red intensity. After balancing out the blue or green color tones, the image appears more natural with colors that are closer to what they would look like outside the waterframe. Thus, after applying this algorithm, the detection model can better distinguish between different parts of the image. Some of the formulas used for calculating average values for the red, green, and blue channels.

20.4.2 Contrast Enhancement Using CLAHE

CLAHE is an acronym for the word Contrast Limited Adaptive Histogram Equalization. It is one of the image enhancement techniques that helps improve the contrast of the images [8] in a smaller section, unlike regular histogram equalization which only focuses on certain sections. It enhances the contrast, making it easier to see important features like marine life or artifacts. Since it focuses on a smaller area, it reveals the detailed features that we might overlook. It also helps to minimize the uneven lighting present in underwater photographs [9]. The best part about this algorithm is that it preserves the original look of the image while improving the contrast.

Figure 20.2 shows the images that have been taken from the dataset, and we have performed image enhancement using the following algorithms.

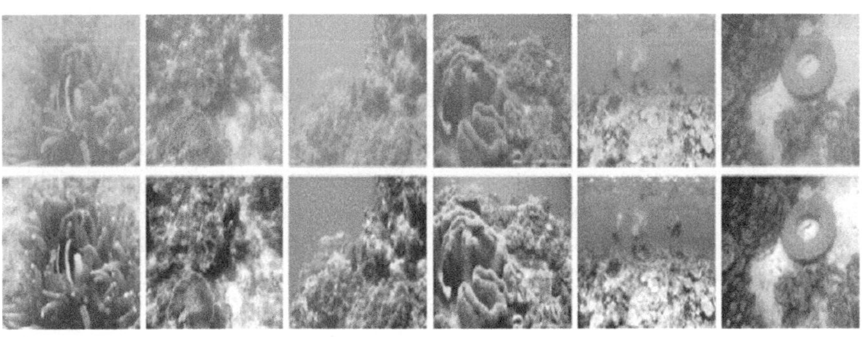

FIGURE 20.2 Original images and enhanced images after applying CLAHE.

20.4.3 Unsharp Masking

The technique sharpens images by increasing the contrast along the edges of objects. It makes the image look more clearer and defined. It also makes the images more appealing, ultimately aiding in better analysis [10] and identification of underwater objects. This is done by first creating a blurred version of the image using a filter called Gaussian blur, which makes the image appear softer, thus reducing the noise. Next, we create a mask by subtracting the blurred image from the original image. This subtraction makes the edges of the objects stand out more. The mask is added to the original image. This process helps to make each and every detail of the image more pronounced and look more sharper. This method also gives the power to control how much sharpening happens by adjusting different settings.

20.4.4 Smoothing Filters

This is the final step of our image enhancement, where we intend to smooth out the variations in an image. It helps to soften the rough edges and reduce small and unwanted details. This makes the image appear more uniform and improves the overall quality. This process helps to achieve clear images, which is essential for accurately identifying objects. This works by averaging the color pixels in a certain area.

20.5 ALGORITHMS USED FOR OBJECT DETECTION

In faster R-CNN, the convolution neural network (CNN) helps extract the features by finding the patterns in the images. The objects that need to be detected often blend into the background underwater due to low contrast and color distortion, making the details unclear. Extracting features helps the model to identify specific parts of an object that make it unique. Features such as edges, textures, shapes, or colors are analyzed, which make marine organisms or artifacts unique from their environment. Edges detect the outline of the object. Textures capture the patterns or the types of surface on the object, which makes it appear rough and smooth. By recognizing these textures the detection system can differentiate easily between different types of objects. Additionally, color is an important feature that helps classify objects easily. Thus, the above features extracted give unique information about an object's identity.

20.5.1 Feature Selection: Region Proposal Network (RPN)

Feature selection is a crucial step in object detection, where the model identifies the most important feature to focus on and ignores the unnecessary ones. This is important for an underwater environment, where a lot of noise or extra information in images due to water particles, plants, or random reflections in the background might confuse the detection system. RPN scans through the image searching for clues where objects could be. Whenever it spots an area that has an object, it marks that area as "potential object location," which denotes that it is the suggested area for the model to look closely at. Then RPN ranks these proposals based on how likely an object can be present. Through this process it filters out the less unusual ones. Only the best proposals further undergo the process of object classification.

20.5.2 Training the Model

Modeling refers to training a Faster R-CNN model to identify and label the objects in the images. The first step in modeling is to gather a large set that includes labeled images. Then the model tries to learn the common characteristics of each object from the given labeled dataset. The models also use mathematical operations called convolution to process the images. Each layer helps the model to break down the images into smaller parts, which enables them to analyze the important features. Training is a crucial step, as it allows the computer to automate the process of identifying and labeling the images. Once the RPN suggests the potential object location, then the model classifies these regions to identify which object is present, such as fish and plastic waste. The model combines the information from features that it has learned with suggested areas to identify and label the objects.

20.5.3 Bounding Box Regression with GIOU Loss

During the prediction phase, bounding boxes are created to point out where an object is located within an image. It gives a clear picture of the position of the object. However, these initial boxes created failed to align perfectly with the object. To solve this problem, we use bounding box regression, which helps to adjust the position and size of each box to get as close as possible to the actual object's boundaries. Generalized Intersection over Union (GIOU) Loss is the measure of the accuracy of this alignment. GIOU measures the distance between the boxes' edges as well as it determines the overlapping area between the predicted and ground-truth boxes. This method enhances the accuracy of the placement and size of bounding boxes and provides more accurate fitting, especially in cases where there are partial overlaps or when the box shape needs adjustment. Thus, using the above process we were able to make boxes closely fit around objects in underwater images.

20.5.4 Soft-NMS

When the model sees multiple boxes for the same object, standard NMS only retains the one with the highest score and eliminates others. Unlike Regular Non-Maximum Suppression, Soft-NMS does not remove the extra boxes but adjusts their confidence scores. This can be a problem, as it shows that the model is unsure about some detection, and ultimately it would result in missing out on detecting required objects just because it had low scores. Soft-NMS, instead of completing or removing the overlapping boxes, retains them but resets their confidence scores based on how much they overlap with the box that has the highest score. Intersection over Union (IoU) calculates the amount of overlap and also determines the area where the two boxes overlap compared to the area covered by both boxes combined.

IoU = Area of Overlap/Area of Union

For each overlapping box, Soft-NMS applies a reduction to its confidence score based on the amount of overlap. Soft-NMS is a better approach than traditional NMS, which helps in managing overlapping detections. It improves accuracy by retaining multiple boxes and adjusting their confidence scores based on how much they overlap with the most confident detection.

20.6 RESULT ANALYSIS

Jupyter Notebook is used to carry out computations and perform data analysis. With the usage of various Python libraries, Jupyter Notebook works with big data sets and represents real-time data visualization through various types of graphs, thus making it a good tool for both exploration and presentation of data. So as we can see, the system is able to execute the task of object detection and recognition using Faster R-CNN.

20.7 CONCLUSION

Underwater image enhancement technique is important for purposes such as object recognition and segmentation of the image, but it is used to show a notable defiance. A new method is established to increase the contrast and quality of the image. Additionally, the upgraded two-step Faster RCNN algorithm is advanced for the improved image enhancement and object detection, which has higher frequency and accuracy in the object detection. The advancement of the method particularly benefits the system in marine research and monitoring of the underwater environment, which helps to find the accurate and improved image quality and detects algorithm which is going to lead to higher precision in objectifying various objects underwater. The enhanced image quality covers the way for the implementation of an updated two-step Faster RCNN and is going to be implemented for the upgraded underwater image enhancement and detection of objects underwater. Furthermore, by applying this innovative algorithm and techniques, we could better learn and preserve our ecosystem underwater, which will ensure that significant marine supplies are going to protect our upcoming generations.

REFERENCES

1. Akkaynak, D., Treibitz, T. 2018. A revised underwater image formation model. *Proceedings of the IEEE Conference on Computer Vision and Pattern Recognition*, pp. 6723–6732.
2. Lin, W.-H. Zhong, J.-X., Liu, S., Li, T., Li, G. 2020. ROIMIX: Proposal-fusion among multiple images for underwater object detection. *IEEE International Conference on Acoustics, Speech and Signal Processing*, pp. 2588–2592.
3. Uplavikar, P.M., Wu, Z., Wang, Z. 2019. All-in-one underwater image enhancement using domain-adversarial learning. *Proceedings of the IEEE Conference on Computer Vision and Pattern Recognition Workshops*, pp. 1–8.

4. Bubeck, S., Sellke, M. 2021. A universal law of robustness via isoperimetry. *Adv. Neural Inform . Proc. Syst.* 34, 2021, pp. 28811–28822.
5. Zhang, J., Zhu, L., Xu, L., Xie, Q. 2020. Research on the correlation between image enhancement and underwater object detection. *Chinese Automation Congress*, 5928–5933.
6. An, S., Xu, L., Senior Member, I., Deng, Z., Zhang, H. 2024. HFM: A hybrid fusion method for underwater image enhancement. *Eng. Appl. Artif. Intell.* 120, 107219.
7. Ancuti, C.O., Ancuti, C., De Vleeschouwer, C., Bekaert, P. 2017. Color balance and fusion for underwater image enhancement. *IEEE Trans. Image Process.* 20 (1), 379–393.
8. Ancuti, C.O., Ancuti, C., De Vleeschouwer, C., Sbert, M. 2019. Color channel compensation (3C): A fundamental pre-processing step for image enhancement. *IEEE Trans. Image Process.* 29, 2053–2065.
9. Anwar, S., Li, C. 2020. Diving deeper into underwater image enhancement: A survey. *Signal Proc. Image Commun.* 89, 115978.
10. Chen, R., Cai, Z., Cao, W. 2021. MFFN: An underwater sensing scene image enhancement method based on multiscale feature fusion network. *IEEE Trans. Geosci. Remote Sens.* 60, 1–12.

21 Explaining and Adapting Self-Healed AI Security Systems for Future Cybersecurity

Sharmila, Dhananjay Kumar, Shashi Bhushan, Priyanka Tyagi, and Mohammad Sadman Tahsin

21.1 INTRODUCTION TO AI IN CYBERSECURITY

Cybersecurity is the practice of protecting systems, networks, and data from unauthorized access, cyber-attacks, and other malicious activities. With the growing use of digital technologies and the internet, cyber threats have become more sophisticated and widespread. Effective cybersecurity ensures the confidentiality, integrity, and availability of data, preventing data breaches, fraud, and other forms of cybercrime that could have serious consequences for individuals, businesses, and governments. As cyber-attacks continue to evolve, traditional defense measures are becoming less effective, and new, more advanced solutions are necessary.

If we thought of cybercrime as an economy, it would be the third largest in the world, just behind the United States and China. In 2021, the global cost of cybercrime was estimated at $6 trillion. Experts predict this cost will continue to rise by about 15% each year, reaching $10.5 trillion annually by 2025 up from $3 trillion in 2015 [1]. This amount not only is much higher than the financial damage caused by natural disasters but also represents the largest economic loss ever seen from a single source. The high cost of cybercrime discourages investment and stifles innovation, as companies and individuals are forced to spend more on security and damage control. Cybersecurity is the practice of protecting technology systems, data, and networks from cyber threats or attacks [2]. A cyber-attack is a crime where someone targets electronic systems to steal, change, or destroy information. Nowadays, cyber-attacks are common, often taking advantage of system weaknesses or human errors to access sensitive information or damage systems. Figure 21.1 shows the top risks under review by ENISA.

- Improving cybersecurity efforts depends much on artificial intelligence (AI). Unlike conventional approaches, AI systems can learn from vast datasets, identify trends, and instantly predict possible hazards, thus offering

DOI: 10.1201/9781003567585-21

FIGURE 21.1 Top risks under review by ENISA.

more dynamic and flexible security measures. Cybersecurity defenses are strengthened using methods including deep learning (DL) and machine learning (ML), thus facilitating quicker identification of security breaches, malware detection, and intrusion prevention [3]. Additionally able to automate some tasks, AI lowers human involvement and increases the general effectiveness of security systems. Though its advantages abound, AI in cybersecurity has certain restrictions [4–8]:

21.2 UNDERSTANDING EXPLAINABLE AI (XAI)

21.2.1 Background of Explainable AI (XAI)

Human-comprehensible artificial intelligence (XAI) represents the collection of methods to demonstrate AI model reasoning to people [12–15]. XAI operates on two levels, which include the specific scenario or prediction explanation at the local level and the entire model explanation at the global level. XAI presents model capabilities and limits along with future system trends and acts as a system for explaining AI model choices through decision-making and prediction mechanisms. XAI refers to several algorithms, tools, and methods meant to provide explanations using visualizing and natural language processing, enabling users to understand and troubleshoot AI systems more efficiently [16]. Making these explanations available helps users to

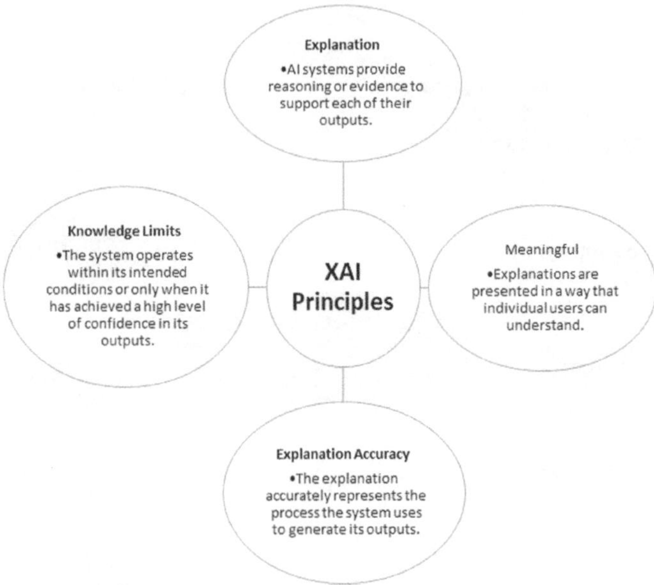

FIGURE 21.2 Principles of XAI by NIST.

understand the learned policies of the AI, which can be examined, interpreted, and changed as necessary without training the system.

Although XAI originated in the 1970s, it has lately attracted a lot of interest, especially with the development of DL. Concerns about their ethical implications, possible biases, and openness have underlined the relevance of XAI as AI systems grow more central in sectors including cybersecurity, finance, and healthcare [17]. Its use spans many spheres, including crop suitability mapping, flood prediction, and drought management, thus influencing important factors including safety, environmental protection, and resource management. As Figure 21.2 shows, the National Institute of Standards and Technology (NIST) proposed four fundamental ideas for explainable artificial intelligence (XAI) systems in mid-2020. A primary distinction is made between interpretability and explainability. Interpretability focuses on understanding the internal cause-and-effect processes of an AI model, while explainability extends beyond this to clarify how and why the model reaches a prediction in a way that is comprehensible to humans [18].

21.2.2 AI Applications and the Black Box Problem

AI has revolutionized industries such as healthcare, manufacturing, finance, and more. AI technologies are embedded in cyber-physical systems (CPS) like automated manufacturing, robotics, healthcare systems, and even social media platforms. While these AI systems offer substantial benefits, their complexity often creates a "black box" problem. The black box issue arises because DL models, known for their high predictive accuracy, are often too complex for humans to easily interpret. The deeper and more complex the model, the harder it becomes to understand how input data is transformed into output decisions, which can hinder trust and accountability in AI systems [19].

While simple models are more interpretable, they lack the sophisticated predictive power of DL models. This trade-off between explainability and accuracy presents a challenge that XAI seeks to address. XAI methods aim to produce transparent, high-performance AI models so that human users may trust and control them efficiently.

21.2.3 Definition and Requirements of XAI Models

XAI is the subset of AI dedicated to create models not only highly performing but also human-understandable ones. It helps users understand and believe in the ML models' decision-making procedures, so guiding AI outputs. As with DL models, these models might be opaque or transparent—that is, intrinsically interpretable. XAI models come in two varieties: post-hoc, in which outside methods are used to explain difficult models following training, ante-hoc, in which the model itself is naturally explainable [20].

XAI models possess distinct key requirements that characterize them according to [21].

- **Comprehensibility**: The model has to present its acquired knowledge in a way humans could grasp.
- **Transparency**: One can grasp a transparent model on their own without outside justifications.
- **Explainability**: The model has to be an interface that approximates the internal AI system processes while yet giving consumers simple, understandable explanations.
- **Understandability**: The user should be able to grasp the model's behavior without needing to delve into its complex algorithmic structure.
- **Accuracy**: XAI models must correctly predict unseen instances, and their performance should be measurable using metrics like accuracy or F1 score.
- **Fidelity**: XAI must accurately represent or approximate the decisions of opaque (black box) models.

XAI guarantees that AI systems remain transparent and efficient by fulfilling these standards, so enabling user confidence and offering management with practical insights for change of AI behavior.

21.2.4 XAI for AI Management

The proper management of AI systems needs a fundamental understanding of XAI systems. AI system monitoring and control together with improvements require a complete understanding of decision-making procedures. XAI enables meaningful human interaction with AI systems, thus supporting compliance with ethical guidelines and human values and regulatory standards [21, 22].

21.3 CYBERSECURITY INTEGRATION OF XAI

21.3.1 Motivation to Integrate XAI into Cybersecurity

The rising volume and sophistication of cyber-attacks including malware, intrusions, and spam have made cybersecurity ever more complicated [23–26]. The enormous volumes of data produced by contemporary technologies including IoT, cloud computing, and edge computing are rendering traditional approaches rule-based, statistical, and signature-based algorithms less effective. These technologies create enormous volumes of network data for which conventional systems find difficulty processing in real time [27–30].

AI in Cybersecurity: The combination of ML together with DL under the AI paradigm functions as an effective solution to manage these sophisticated cybersecurity issues. Virtual intelligence systems perform three functions in cybersecurity including malware identification and network traffic evaluation and anomaly detection work. However, AI models in cybersecurity face significant challenges [31]:

- **Lack of Transparency:** AI models, particularly DL models, often operate as "black boxes." This implies that even if they could be quite accurate, humans cannot easily follow their decision-making procedures.
- **Vulnerability to Adversarial Attacks:** Adversarial inputs allow one to manipulate AI models, so fooling them into making bad decisions.
- **Privacy and Ethical Concerns:** AI integration into cybersecurity presents privacy concerns since many AI models depend on access to vast volumes of private data to operate as expected.

The requirement for XAI emerges from the necessity to reduce security threats that occur when depending exclusively on AI for cybersecurity systems. XAI functions to make AI systems explainable through transparency, which enables humans to understand their decision-making processes [32–33]. Figure 21.3 shows the classification of XAI.

21.3.2 Core Components of Self-Healing AI in Cybersecurity

The core components and working of self-healing AI in cybersecurity are as follows [42–50]:

(A) **Anomaly Detection and Threat Identification**

Self-healing AI systems use anomaly detection models trained on typical network behavior to identify irregular patterns in network traffic, user actions, and application processes [51]. These models build a baseline of normal activity by using ML and DL techniques [34–36] including clustering and supervised or reinforcement learning.

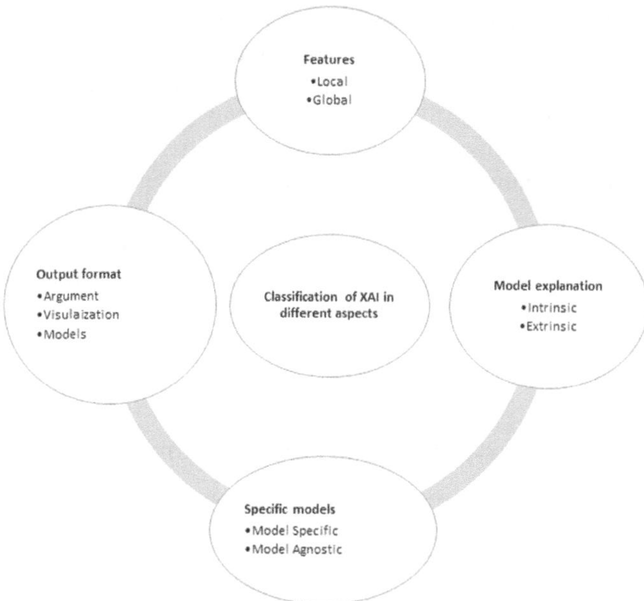

FIGURE 21.3 Classification of XAI.

Deviations from this baseline signal potential threats, allowing the system to take preventive or corrective actions [52].

(B) Real-Time Monitoring and Data Analysis

Continuous monitoring across network activities and system logs enables self-healing AI to analyze extensive datasets in real time, detecting security issues as they emerge. Natural language processing and real-time stream processing are among the advanced data analysis tools that let the AI spot indicators of known and unknown risks. Preventive and timely threat management depends on this real-time capacity [53].

(C) Automated Diagnosis and Vulnerability Assessment

The self-healing mechanism begins evaluation when it identifies potential threats by assessing their source together with their impact area [37–38] and severity level. Through evaluated data from known vulnerabilities and historic trends, the system determines its points of attack along with areas that need urgent attention. This evaluation procedure decides which course of action to take, ranging from minor system changes to complete system segregation if needed.

(D) Automated Patching and Recovery

Trained for self-healing AI takes over distributing updates or configuring changes to guard security vulnerabilities so reducing possible exploits. The system implements transitory protection strategies or switches to previous stable states when zero-day vulnerabilities exist since permanent solutions require development time. The automated patching system both speeds up threat elimination while reducing the need for manual human actions.

(E) **Adaptive Learning and Evolution**

The primary characteristic of self-healing AI originates from its reinforcement learning ability that allows the system to learn during past events. While it learns better about threat pathways to create enhanced predictive capabilities, the AI gets stronger in resisting such assaults. The system's on-going learning abilities enable its operation against growing cyber threats.

21.3.3 Future of Self-Healing AI in Cybersecurity

Self-healing systems will probably find integration across cybersecurity systems as AI develops, providing layered, adaptive defenses that change with new risks. Including Explainable AI (XAI) into self-healing systems would improve openness even more and help cybersecurity teams to grasp the autonomous decisions and actions of the system. This junction of human supervision and self-healing AI promises to produce strong, flexible cybersecurity systems guaranteeing data integrity and privacy in ever more complicated digital environments.

21.4 INTEGRATION OF XAI FRAMEWORK FOR CYBERSECURITY

By offering understandable and actionable insights into how AI systems identify and react to threats, XAI systems in cybersecurity seek to improve the openness and trustworthiness of AI-driven security solutions [39–41]. In cybersecurity especially, where decision-makers must guarantee accurate threat detection and efficient response by clearly explaining automated actions [51]. This framework acts as a general workflow with variations depending on the particular cybersecurity activity being addressed, as depicted in Figure 21.4.

21.5 DATASET OF CYBERSECURITY

There is great importance in how data selection and application support the advancement of cybersecurity scientific research. The effectiveness of XAI models depends on both the quality and quantity of collected data. regardless of whether they originate from DL or ML systems. The methods for collecting cybersecurity data include software tools like WinDump or Wireshark, but these techniques work best with modest datasets unless used for their intended scope. Furthermore, they are costly and time-consuming. Using benchmark cybersecurity datasets provides a more effective alternative to help overcome these difficulties, so saving time and money related to data collection and improving the validity of research. These datasets let academics evaluate, validate, and train XAI-based cybersecurity solutions. Table 21.1 shows the datasets available for cyber-attacks.

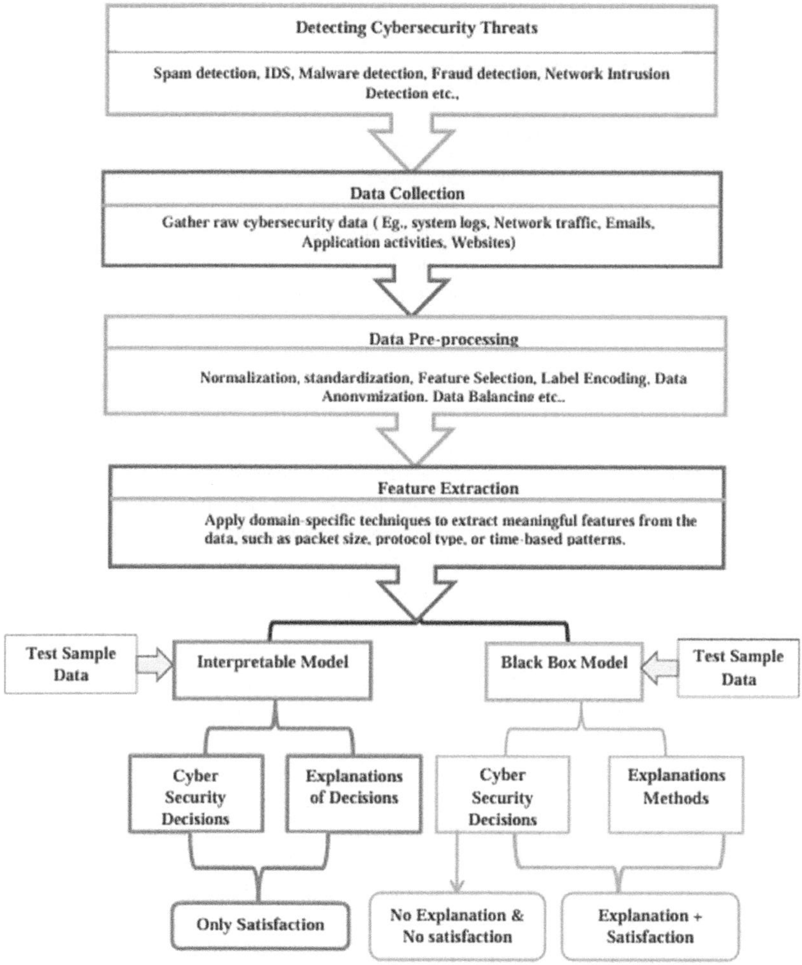

FIGURE 21.4 XAI Framework for cybersecurity.

21.6 EXPLAINABLE AI TO PROTECT AGAINST CYBER ATTACKS

21.6.1 Malware Attack

XAI is increasingly crucial in cybersecurity for enhancing the interpretability of AI models, particularly in malware detection. Malware represents a significant threat on the Internet, and the vast volume of malware necessitates efficient detection methods. Traditional malware detection techniques fall into two categories: static and dynamic. Static detection analyzes malware binaries without execution, which can be circumvented by techniques such as code obfuscation. Dynamic detection, on the other hand, involves executing malware in a controlled environment to observe its behavior. While effective, these methods are resource-intensive

TABLE 21.1
Dataset Available for Cyber Attack

Attack Category	Dataset Name	Year	Dataset Details
Malware	BIoT [54]	2018	BIoT dataset captures network traffic from a range of IoT devices infected by botnets. This dataset, primarily for malware analysis, includes details on DDoS attacks initiated by these infected devices.
	IoTPOT [55]	2016	IoTPOT dataset contains malware samples targeting IoT devices, offering valuable insight into malware behavior in diverse CPU environments.
	IoT-23 [56]	2020	IoT-23 captures data from 23 types of IoT malware. It focuses on traffic details, aiding in the study of malware communication patterns and device infection pathways.
	EMBER [57]	2018	The EMBER dataset consists of a broad set of malware features useful for ML, providing both training and validation data for malware detection models.
	Genome Project [58]	2012	This extensive dataset offers malware samples collected from various sources, providing insight into malware signatures and assisting in the classification of malicious software.
	VirusShare Update [59]	2015	VirusShare dataset includes a wide array of malware samples and is continually updated, making it a reliable source for malware behavior analysis.
	CC-Amy [60]	2018	This collection includes labeled samples of malware targeting Android systems, providing insights into mobile malware threats and detection methodologies.
	DREBIN [61]	2014	DREBIN is designed for mobile security research, including extensive labeled malware data targeting Android devices.
Spam	SMS Spam [62]	2017	SMS Spam dataset provides a collection of spam and non-spam SMS messages, useful for text classification and spam detection studies.
	Enron [63]	2004	The Enron dataset includes a vast repository of email data, valuable for training models in spam filtering and email classification.
	ISCX-URL2016 [64]	2016	Comprising over 400,000 labeled URLs, this dataset focuses on categorizing URLs as benign or malicious, aiding in the development of web security applications.
	NSL-KDD [65]	2009	NSL-KDD dataset refines the original KDD'99, reducing redundancy and complexity while focusing on both spam and network intrusion detection.
	UNSW-NB15 [66]	2015	UNSW-NB15 includes modern network traffic data for intrusion detection research, with various network attacks represented to assist in anomaly detection.

(*Continued*)

TABLE 21.1 (CONTINUED)
Dataset Available for Cyber Attack

Attack Category	Dataset Name	Year	Dataset Details
Network intrusion	AWID [67]	2016	AWID captures wireless network traffic from 10 devices, specifically designed to study intrusion detection in small wireless environments.
	CIC-IDS2017 [68]	2017	CIC-IDS2017 comprises diverse attack patterns, including DoS, DDoS, and other network intrusions, facilitating research into network defense mechanisms.
	CIC-DDoS2019 [69]	2019	CIC-DDoS2019 dataset offers detailed instances of DDoS attacks, useful for understanding patterns in large-scale denial-of-service attacks.
	TON_IoT [70]	2020	This dataset covers IoT data alongside cybersecurity data, offering a mix of normal and attack traffic to help develop IoT-specific security solutions.
	ADFA-LD [71]	2013	ADFA-LD dataset focuses on host-based intrusion detection, providing traces from typical operating system logs under both attack and normal conditions.
Botnet	ISOT Botnet [72]	2011	ISOT Botnet dataset contains labeled botnet traffic, beneficial for studying command and control channels in botnet networks.
	Bot-IoT [73]	2019	This dataset contains extensive IoT botnet activity, designed for use in the analysis of IoT-specific botnet attacks such as Mirai.
DGA	UMUDGA [74]	2020	UMUDGA provides domain generation algorithm (DGA) data used by botnets, useful for studying patterns in DGA-based malicious domains.
Phishing	ISCX-URL2016 [75]	2016	This URL-based dataset categorizes phishing and legitimate URLs, aiding in web security research to detect and prevent phishing attempts.
CPSs	SWaT [76]	2014	The SWaT dataset, designed for cyber-physical systems, captures real-world water treatment plant data for detecting anomalies and intrusions.

and slow when applied manually across extensive datasets. To improve efficiency, AI-based approaches, particularly DL models, have become common in malware detection due to their superior performance over traditional methods. However, the opaque nature of neural networks—often referred to as "black boxes"—limits their transparency. As a result, researchers are integrating XAI techniques to make these AI-based systems more interpretable and trustworthy, especially when deployed in new environments [77].

21.6.2 SPAM

The rise in internet usage has led to an increase in spam, with over 216.4 billion emails sent daily in 2021, of which more than 55% were spam [78]. AI systems, particularly those that can evolve and adjust themselves, have been proposed as effective solutions to combat spam. However, concerns about privacy and the legal complexities of spam often raise questions about AI's decision-making process, especially regarding black-box ML and DL models [79]. Common inquiries about spam classification mechanisms and how AI determines legitimate from spam messages occur often [80].

Botnets are networks of compromised computers that carry out coordinated malicious activities, such as disrupting resources and crashing websites. The botnet lifecycle includes five phases: Initial Injection, Secondary Injection, Connection, Malicious Activities, and Maintenance and Updating [81]. Due to the significant threat botnets pose, the botnet detection market is expected to grow rapidly, with projections of reaching $965.6 million by 2027 at a compound annual growth rate (CAGR) of 24%. XAI enhances botnet detection by adding transparency, which helps users understand and trust the outputs of automated systems, thus reducing automation bias. Hatma et al. [83] proposed a model for botnet Domain Generation Algorithm (DGA) detection that combined SHAP and LIME explanations with five ML algorithms, with Random Forest achieving the highest accuracy (96.3%) [82]. Despite this, the model faced challenges in handling Mask botnet attacks and temporal complexity in feature calculation. To categorize botnets, Shohei et al. [84] developed a two-step clustering approach using DBSCAN and frequent pattern mining on two datasets, MAWI and ISP, which successfully identified bot groups from a large number of IP addresses. The results were further visualized with a dendrogram, providing an interpretable structure for understanding botnet clusters. This visualization technique enhanced the interpretation of botnet detection results. Michele et al. [85] introduced ReTweet-Tweet (RTT), a scatterplot to help visualize retweet patterns of user accounts. Their detection system RTbust, based on Variational Autoencoders (VAEs) and Long Short-Term Memory (LSTM) networks, uses RTT to give insights into accounts classified as bots. For feature reduction, Prinent Analysis (PCA) has been used to simplify botnet detection models, as demonstrated by Hayretdin et al. [86], who used Decision Trees to maintain interpretability. The Decision Tree classifier enabled security analysts to quickly understand the features that led to each classification. In DL applications, Partha et al. [87].

21.6.3 Phishing Attack

Phishing, which involves fake messages designed to steal sensitive data or install malware, is a prevalent type of online fraud [88]. Explainable AI (XAI) has been instrumental in enhancing the detection and understanding of phishing threats through several innovative approaches. Multi-Modal Hierarchical Attention Model (MMHAM) by Yidong et al. [89] proposed a model that detects phishing websites by analyzing three modalities of website content: URLs, text, and images. The system self-explains by identifying content areas that most contribute to classification results, achieving an accuracy of 91.2% and an F1-score of 90.7% with the use of distinct feature filters. A model for URL-based phishing detection was developed using the Ebbu2017 dataset. The approach employed LIME and Explainable Boosting Machine (EBM) techniques to interpret phishing predictions, while Random Forest and SVM classifiers were used for evaluation [90].

21.6.4 Network Intrusion

Network intrusions refer to unauthorized access to a computer or network within a domain [91]. To counteract such threats, Network Intrusion Detection Systems (NIDS) are implemented, which monitor system activity for signs of malicious behavior. The use of ML and DL algorithms in NIDS has grown significantly, along with the integration of Explainable AI (XAI) to improve the robustness and interpretability of these systems [92].

Several studies have focused on improving NIDS efficiency with explainability:

- Pieter et al. [93] developed a two-stage pipeline using XGBoost and Autoencoder, incorporating the SHAP method to explain the model's decisions. On the NSL-KDD dataset, the method improved current models in accuracy and precision.
- Giuseppina et al.'s [94] ROULETTE presented an explainable neural network for network traffic classification, using attention layers for improved interpretability, demonstrated using heatmaps for feature relevance.
- Using LIME, SHAP, and RuleFit for explanations, Zakaria et al. [95] proposed a DL-XAI hybrid system for IoT intrusion detection, so improving the understanding of IDS decisions.
- While cutting convolution layers for better explainability without sacrificing accuracy, Yiwen et al. [96] merged neural networks and tree models to detect DDoS and flood attacks.
- Sivamohan et al. [97] introduced BiLSTM-XAI with Krill Herd Optimization (KHO) to enhance model explainability and performance on NSL-KDD and Honeypot datasets.
- Hong et al. [98] developed the FAIXID framework, using multiple XAI techniques to enhance decision transparency and help analysts mitigate false positives.
- Shraddha et al. [99] presented a system using various XAI methods to provide detailed explanations of intrusion detection decisions, improving model transparency.

The Decision Tree plays an important role in intrusion detection systems by enabling feature ranking and rule extraction, while P-based Random Forest systems also detect malicious network traffic intrusion. A system for DNS over HTTPS traffic anomaly detection, which Additionina et al. [100] developed, shows both high accuracy and effective interpretability.

21.6.5 DOS Attack

A major threat to the Internet is denial-of-service (DoS) attacks, whereby malevolent nodes create false messages meant to cause network traffic disturbance or resource depletion of other nodes. Traditional Intrusion Detection Systems (IDS) find it difficult to identify these sophisticated attacks as DoS attacks change and get more complicated because of their ever more complex patterns. ML and DL models have been used to more successfully identify DoS attacks in order to handle this difficulty.

Apart from these models, XAI approaches have been applied to improve openness and comprehension of the predictions of these models by means of feature-based investigations on algorithm-based decisions. Using CICFlowMeter for flow extraction and feature extraction to enhance the performance of ML models for DoS attack detection, Boryau et al. [101] presented CSTITool. By converting packet data into flow data, CICFlowMeter reduces the data size and so lessens the demand for storage. To build an XGBoost model, the study made use of several attack data including Remote Access Software from the CIC-IDS2017 dataset and the CSTI-10 malware dataset as well as Network Service Scanning, Endpoint DoS, Brute Force [102]. With Precision and Recall improving by 1.23% and 1.59%, respectively, using additional flow statistics from CSTITool proved to enhance the performance of the detection model [103–110]. Furthermore, the relationship between cyber-attacks and network flow variables was investigated using the XAI approach SHAP (Shapley Additive Explanations), offering further understanding of the decision-making process of the model [111–114].

21.6.6 XAI in Cybersecurity for Smart Financial Systems

Through AI models the financial sector experienced a paradigm shift that brought lower expenses and better operational performance to fields such as asset management and investment advice and risk prediction and lending operations as well as improved customer interactions [115]. While these AI-driven systems enhance efficiency, they also increase the vulnerability to cyberattacks, as financial systems remain prime targets due to their value. To protect against cybercrime in smart financial systems, one of the key measures should be the integration of Explainable AI (XAI) [116]. This is especially important in high-stakes areas like money laundering detection and corporate mergers and acquisitions, where not only accuracy and robustness are necessary but also the ability to provide understandable justifications to gain users' trust in automated decisions.

Swati et al. [117] introduced a belief-rule-based (BRB) automated AI decision-support system for loan underwriting. This system incorporates human expertise and can apply supervised learning to extract knowledge from previous data. BRB's hierarchical structure accommodates both factual and heuristic rules. The rules triggered by data points representing loan applications, as well as the contribution of specific attributes in activated rules, help explain the decision-making process. Verification of administered loan rejection information originates from this system for transparent responses to applicants.

Within a corporate mergers and acquisitions (M&A) framework, the authors offered a novel approach to generate consistent counterfactual explanations [118]. To increase model resistance, the method makes advantage of an adversarial training technique applied in a transformer-based classifier. A masked language model that enhanced word significance measurement for more consistent counterfactual explanation generation attracted interest on financial text categorization. In terms of accuracy as well as the capacity to create explanations, the proposed approach exceeded conventional models such as SVM, CNN, BiGRU, and HAN.

Ambreen et al. [119] constructed an evidence-based interactive system to improve customer trust and comprehension of AI algorithm results in the smart banking sector. This approach involved creating a digital dashboard to facilitate engagement with algorithm outputs and boost data scientists' confidence in their understanding of AI-driven decisions. In this model, a probabilistic neural network (PNN) was used to classify various bank transaction scenarios.

21.6.7 XAI in Cybersecurity for Transportation

The integration of advanced technologies like software-defined networks (SDNs), Industrial Internet of Things (IIoT), blockchain, AI, and vehicular ad hoc networks (VANETs) has enhanced the operational capabilities of smart transportation systems. However, these innovations have also increased vulnerability to security breaches, exposing transportation systems to potential intrusions. Security issues in transportation technologies, such as Wireless Sensor Networks (WSN), Vehicle-to-Everything (V2X) communication, Variable Message Signs (VMS), and Traffic Signal Controllers (TSC), have already made them targets of cyberattacks or left them susceptible to future threats. To mitigate these risks, AI-driven intrusion detection systems have been introduced. Despite notable advancements in AI's role within smart transportation systems, explainable AI (XAI) methods are still necessary. XAI can enhance the monitoring and understanding of various transportation factors, such as driver behavior, accident causality, and vehicle conditions.

A ML approach for detecting misbehaving vehicles within VANETs was proposed by Harsh et al., where the system's stability and safety depend on the accuracy of information shared between autonomous vehicles (AVs). Misinformation from misbehaving vehicles can jeopardize the entire network, making their detection essential for maintaining a robust VANET. The Vehicular Reference Misbehavior (VeReMi) dataset was used in an ensemble learning framework, leveraging Random Forest and decision tree-based algorithms to achieve high accuracy (98.43%) and F1 score (98.5%).

Shideh et al. [120] introduced a transportation energy model (TEM) using the XAI technique LIME, which forecasts household transportation energy usage based on data from the Household Travel Survey (HTS). The model achieved an impressive validation accuracy of 83.4%, with the XAI framework enabling both local explanation of inputs (such as household travel and neighborhood data) and an assessment of the model's confidence in its predictions.

Bustos et al. [121] proposed an automated system for reducing traffic-related fatalities using computer vision techniques, such as image segmentation and class activation mapping. The approach trains a Residual Convolutional Neural Network to generate a risk index for urban environments, providing a detailed map of risk levels and potential safety measures for both pedestrians and vehicles.

21.7 CONCLUSION

This chapter has shown how XAI is revolutionizing cybersecurity by means of more understandable and trustworthy AI-driven tools. From malware and botnets to phishing and denial-of-service attacks, as cyber threats get more complicated using AI in security has become crucial. Traditional AI systems, on the other hand, sometimes operate like a "black box," in which case users cannot see or comprehend how these systems arrive at their decisions. Lack of openness might cause mistrust and missed insights. XAI clarifies how artificial intelligence makes decisions, thereby addressing this issue.. XAI tools, for instance, can show which bits of data such as odd login behavior in phishing attempts or suspicious file patterns in malware most support the alerts of a system. Because cybersecurity teams understand alert sources, their enhanced transparency helps them to respond with accuracy. Visualization tools including feature maps and diagrams enable cybersecurity teams to detect data relationships, thereby improving their identification and defense against threats. XAI application in cybersecurity raises many challenges. Three key issues for XAI models must be handled by security experts: they need consistent performance evaluations, high-quality data sources, and protection from attackers trying to fool models. The fixing of these issues calls for constant cooperation between government legislators and developer security experts. Establishing trust between users and AI security systems depends on XAI, which is therefore indispensable. Built-in explainability features of XAI systems enable security tools to deliver advanced functionalities together with easy user interfaces, helping organizations protect themselves from cyber threats more effectively but with understandable and consistent solutions. The progress of this discipline guarantees that XAI will be indispensable for creating digital environments combining resilience with security.

REFERENCES

1. Morgan, S. "Special report: Cyberwarfare in the C-suite, 1666 Online." *Cybercrime Magazine*, 2020. Available online at https://cybersecurity1667ventures.com/cybercrime-damages-6-trillion-by-2021/.
2. ENISA. *Enisa Threat Landscape 2020—List of Top 15 Threats.* Athens, Greece: ENISA, 2020.
3. Vilone, Giulia, and Luca Longo. "Explainable artificial intelligence: A systematic review." 2020. *arXiv preprint arXiv:*2006.00093.
4. Speith, Timo. "A review of taxonomies of explainable artificial intelligence (XAI) methods." In *2022 ACM Conference on Fairness, Accountability, and Transparency*, 2022.

5. Nazar, M., M. M. Alam, E. Yafi, and M. S. Mazliham. "A systematic review of human–computer interaction and explainable artificial intelligence in healthcare with artificial intelligence techniques." *IEEE Access* 9 (2021): 153316–153348,
6. Pradhan, Romila, et al. "Explainable AI: Foundations, applications, opportunities for data management research." In *2022 IEEE 38th International Conference on Data Engineering, ICDE*. IEEE, 2022.
7. Vilone, G., and L. Longo. "Classification of explainable artificial intelligence methods through their output formats." *Machine Learning and Knowledge Extraction* 3 (2021): 615–661. https://doi.org/10.3390/make2121032.
8. Angelov, Plamen, et al. "Explainable artificial intelligence: An analytical review." *Wiley Interdisciplinary Reviews: Data Mining and Knowledge Discovery* 11, no. 5 (2021): e1424.
9. Islam, M. R., M. U. Ahmed, S. Barua, and S. Begum. "A systematic review of explainable artificial intelligence in terms of different application domains and tasks." *Applied Sciences* 12 (2022): 1353. https://doi.org/10.3390/app12031353.
10. Han, M., and J. Kim. "Joint banknote recognition and counterfeit detection using explainable artificial intelligence." *Sensors* 19 (2019): 3607. https://doi.org/10.3390/s19163607.
11. Loh, Hui Wen, Chui Ping Ooi, Silvia Seoni, Prabal Datta Barua, Filippo Molinari, and U. Rajendra Acharya. "Application of explainable artificial intelligence for health care: A systematic review of the last decade (2011–2022)." *Computers in Methods and Programs in Biomedicine* 226 (2022). https://doi.org/10.1016/j.cmpb.2022.107161.
12. Zhang, Y., Y. Weng, and J. Lund. "Applications of explainable artificial intelligence in diagnosis and surgery." *Diagnostics* 12 (2022): 237. https://doi.org/10.3390/diagnostics12020237.
13. Keneni, B. M., et al. "Evolving rule-based explainable artificial intelligence for unmanned aerial vehicles." *IEEE Access* 7 (2019): 17001–17016. https://doi.org/10.1109/ACCESS.2019.2893141.
14. Bahani, Khalid, Mohammed Moujabbir, and Mohammed Ramdani. "An accurate fuzzy rule-based classification system for heart disease diagnosis." *Scientific African* 14 (2021): e01019.
15. Linardatos, P., V. Papastefanopoulos, and S. Kotsiantis. "Explainable AI: A review of machine learning interpretability methods." *Entropy* 23 (2021): 18. https://doi.org/10.3390/e22110018.
16. Ayyar, M. P., J. Benois-Pineau, and A. Zemmari. "Review of white box methods for explanations of convolutional neural networks in image classification tasks." *Journal of Electronic Imaging* 21, no. 5 (2021): 050901.
17. Burkart, Nadia, and Marco F. Huber. "A survey on the explain ability of supervised machine learning." *Journal of Artificial Intelligence Research* 70 (2021): 245–317.
18. Zafar, Muhammad Rehman, and Naimul Mefraz Khan. "DLIME: A deterministic local interpretable model-agnostic explanations approach for computer-aided diagnosis systems." 2019. *arXiv preprint arXiv:*1906.10263.
19. Sahakyan, Maria, Zeyar Aung, and Talal Rahwan. "Explainable artificial intelligence for tabular data: A survey." *IEEE Access* 9 (2021): 135392–135422.
20. Palatnik de Sousa, I., M. Maria Bernardes Rebuzzi Vellasco, and E. Costa da Silva. "Local interpretable model-agnostic explanations for classification of lymph node metastases." *Sensors* 19 (2019): 2169. https://doi.org/10.3390/s19132169.
21. Guidotti, Riccardo, Anna Monreale, Salvatore Ruggieri, Franco Turini, Fosca Giannotti, and Dino Pedreschi. "A survey of methods for explaining black box models." *ACM Computing Surveys* 51, no. 5 (2018): 93, (2019): 42. https://doi.org/10.1145/3236009.

22. Van den Broeck, Guy, et al. "On the tractability of SHAP explanations." *Journal of Artificial Intelligence Research* 74 (2022): 851–886.
23. Ponn, Thomas, Thomas Kröger, and Frank Diermeyer. "Identification and explanation of challenging conditions for camera-based object detection of automated vehicles." *Sensors* 20, no. 13 (2020): 3699.
24. Lombardi, Angela, et al. "Explainable deep learning for personalized age prediction with brain morphology." *Frontiers in Neuroscience*, 2021 (2021): 578.
25. Utkin, Lev, and Andrei Konstantinov. "Ensembles of Random SHAPs." *Algorithms* 15, no. 11 (2022): 431.
26. Adadi, A., and M. Berrada. "Peeking Inside the Black-Box: A Survey on Explainable Artificial Intelligence (XAI)." *IEEE Access* 6 (2018): 52138–52160. https://doi.org/10.1109/ACCESS.2018.2870052.
27. Sheh, R., and I. Monteath. "Defining explainable AI for requirements analysis." *Künstliche Intelligenz* 32 (2018): 261–266. https://doi.org/10.1007/s13218-018-0559-3.
28. Phillips, P. Jonathon, et al. *Four Principles of Explainable Artificial Intelligence*. Gaithersburg, Maryland, 2020.
29. Gunning, D. W., and D. Aha. "DARPA's explainable artificial intelligence program." *AI Magazine* 40, no. 2 (2019): 44.
30. Wolf, C. T. "Explainability scenarios: Towards scenario-based XAI design." In *Proceedings of the 24th International Conference on Intelligent User Interfaces*, New York, NY, USA, March 2019, 252–257. https://doi.org/10.1145/3211275.3212317.
31. Barredo Arrieta, A., N. Díaz-Rodríguez, J. Del Ser, A. Bennetot, S. Tabik, A. Barbado, S. Garcia, S. Gil-Lopez, D. Molina, R. Benjamins, R. Chatila, and F. Herrera. "Explainable Artificial Intelligence (XAI): Concepts, taxonomies, opportunities, and challenges toward responsible AI." *Information Fusion* 58 (2020): 82–115. https://doi.org/10.1016/j.inffus.2019.12.012.
32. Wolf, C. T. "Explainability Scenarios: Towards scenario-based XAI Design." In *Proceedings of the 24th International Conference on Intelligent User Interfaces*, New York, NY, USA, March 2019, 252–257. https://doi.org/10.1145/3211275.3212317.
33. Barredo Arrieta, A., N. Díaz-Rodríguez, J. Del Ser, A. Bennetot, S. Tabik, A. Barbado, S. Garcia, S. Gil-Lopez, D. Molina, R. Benjamins, R. Chatila, and F. Herrera. "Explainable Artificial Intelligence (XAI): Concepts, taxonomies, opportunities, and challenges toward responsible AI." *Information Fusion* 58 (2020): 82–115. https://doi.org/10.1016/j.inffus.2019.12.012.
34. Ying, R., D. Bourgeois, J. You, M. Zitnik, and J. Leskovec. "GNNExplainer: Generating explanations for graph neural networks." 2019. *arXiv* 1903.03894.
35. Lundberg, S. M., and S.-I. Lee. "A unified approach to interpreting model predictions." In *Proceedings of the Advances in Neural Information Processing Systems* 30 (2017): 1–10. https://proceedings.neurips.cc/paper/2017/hash/8a20a8621978632d76c43dfd28b67767-Abstract.html.
36. Iyer, R., Y. Li, H. Li, M. Lewis, R. Sundar, and K. Sycara. "Transparency and explanation in deep reinforcement learning neural networks." 2018. *arXiv* 1809.06061.
37. Selvaraju, R. R., M. Cogswell, A. Das, R. Vedantam, D. Parikh, and D. Batra. "Grad-CAM: Visual explanations from deep networks via gradient-based localization." *International Journal of Computer Vision* 128, no. 2 (2020): 336–359. https://doi.org/10.1007/s11263-019-01228-7.
38. Wachter, S., B. Mittelstadt, and C. Russell. *Counterfactual explanations without opening the black box: Automated decisions and the GDPR*. Tech. Rep. 2017. https://doi.org/10.2139/SSRN.3063289.
39. Ibrahim, M., M. Louie, C. Modarres, and J. Paisley. "Global explanations of neural networks: Mapping the landscape of predictions." 2019. *arXiv* 1902.02384.

40. Liu, H., Q. Yin, and W. Yang Wang. "Towards explainable NLP: A generative explanation framework for text classification." 2018. *arXiv* 1811.00196.
41. Danilevsky, M., K. Qian, R. Aharonov, Y. Katsis, B. Kawas, and P. Sen. "A survey of the state of explainable AI for natural language processing." 2020. arXiv 2010.00711.
42. Jeyakumar, J. V., J. Noor, Y.-H. Cheng, L. Garcia, and M. Srivastava. "How can i explain this to you? An empirical study of deep neural network explanation methods." In *Proceedings of the Advances in Neural Information Processing Systems* 33 (2020): 4211–4222. https://proceedings.neurips.cc/paper/2020/hash/2c29d89cc56cdb191c60db2f0bae796b-Abstract.html.
43. Jin, W., X. Li, and G. Hamarneh. "Evaluating explainable AI on a multimodal medical imaging task: Can existing algorithms fulfill clinical requirements?" 2022. *arXiv* 2203.06487.
44. Lu, J., D. Lee, T. W. Kim, and D. Danks. "Good explanation for algorithmic transparency." Rochester, NY, USA. Nov. 2019. https://doi.org/10.2139/SSRN.3503603.
45. Amgoud, L., and H. Prade. "Using arguments for making and explaining decisions." *Artificial Intelligence* 173, no. 3–4 (2009): 413–436. https://doi.org/10.1016/j.artint.2008.11.006.
46. Wu, M., M. Hughes, S. Parbhoo, M. Zazzi, V. Roth, and F. Doshi-Velez. "Beyond sparsity: Tree regularization of deep models for interpretability." In *Proceedings of the AAAI Conference on Artificial Intelligence* 32, no. 1 (2018): 1–8. https://doi.org/10.1609/aaai.v32i1.11501.
47. Lakkaraju, H., E. Kamar, R. Caruana, and J. Leskovec. "Faithful and customizable explanations of black box models." In *Proceedings of the AAAI/ACM Conference on AI, Ethics, and Society*, New York, NY, USA, 2019, 131–138. https://doi.org/10.1145/3306618.3314229.
48. Fidel, G., R. Bitton, and A. Shabtai. "When explainability meets adversarial learning: Detecting adversarial examples using SHAP signatures." 2019. *arXiv* 1909.03418.
49. Guo, W. "Explainable artificial intelligence for 6G: Improving trust between human and machine." *IEEE Communications Magazine* 58, no. 6 (2020): 39–45. https://doi.org/10.1109/MCOM.001.2000050.
50. Hussain, F., R. Hussain, and E. Hossain. "Explainable Artificial Intelligence (XAI): An engineering perspective." 2021. *arXiv* 2101.03613.
51. Xin, Y., L. Kong, Z. Liu, Y. Chen, Y. Li, H. Zhu, M. Gao, H. Hou, and C. Wang. "Machine learning and deep learning methods for cybersecurity." *IEEE Access* 6 (2018): 35365–35381. https://doi.org/10.1109/ACCESS.2018.2836950.
52. Yavanoglu, O., and M. Aydos. "A review on cyber security datasets for machine learning algorithms." In *Proc. IEEE Int. Conf. Big Data (Big Data)* (2017): 2186–2193. https://doi.org/10.1109/BigData.2017.8258167.
53. Xin, Y., L. Kong, Z. Liu, Y. Chen, Y. Li, H. Zhu, M. Gao, H. Hou, and C. Wang. "Machine learning and deep learning methods for cybersecurity." *IEEE Access* 6 (2018): 35365–35381. https://doi.org/10.1109/ACCESS.2018.2836950.
54. Y. Meidan, M. Bohadana, Y. Mathov, Y. Mirsky, A. Shabtai, D. Breitenbacher, and Y. Elovici. "N-baiot-network-based detection of IoT botnet attacks using deep autoencoders." *IEEE Pervasive Computing* 17, no. 3 (2018): 12–22. https://doi.org/10.1109/MPRV.2018.03367731.
55. Y. M. P. Pa, S. Suzuki, K. Yoshioka, T. Matsumoto, T. Kasama, and C. Rossow. "IoTPOT: A novel honeypot for revealing current IoT threats." *Journal of Information Processing* 24, no. 3 (2016): 522–533. https://doi.org/10.2197/ipsjjip.24.522.
56. S. Garcia, A. Parmisano, and M. J. Erquiaga. "IoT-23: A labeled dataset with malicious and benign IoT network traffic." *Zenodo, Technical Report* (2020). https://doi.org/10.5281/zenodo.4743746.

57. H. S. Anderson, and P. Roth. "EMBER: An open dataset for training static PE malware machine learning models." 2018. *arXiv*:1804.04637.
58. Y. Zhou, and X. Jiang. "Dissecting Android malware: Characterization and evolution." In *Proceedings of the IEEE Symposium on Security and Privacy* (May 2012): 95–109. https://doi.org/10.1109/SP.2012.16. VirusShare.com, Accessed: Jul. 13, 2022. [Online]. Available: https://virusshare.com/
59. A. H. Lashkari, A. F. A. Kadir, L. Taheri, and A. A. Ghorbani. "Toward developing a systematic approach to generate benchmark Android malware datasets and classification." In *Proceedings of the International Carnahan Conference on Security Technology (ICCST)* (Oct. 2018): 1–7. https://doi.org/10.1109/CCST.2018.8585560.
60. D. Arp, M. Spreitzenbarth, M. Hubner, H. Gascon, K. Rieck, and C. E. R. T. Siemens. "DREBIN: Effective and explainable detection of Android malware in your pocket." In *Proceedings NDSS*, 2014, 23–26. [Online]. https://www.ndss-symposium.org/ndss2014/programme/drebin-effective-and-explainable-detection-android-malware-your-pocket/
61. T. A. Almeida, J. M. G. Hidalgo, and A. Yamakami. "Contributions to the study of SMS spam filtering: New collection and results." In *Proceedings 11th ACM Symp. Document Eng.,* New York, NY, USA, 2011, 259–262. https://doi.org/10.1145/2034691.2034742.
62. V. Metsis, I. Androutsopoulos, and G. Paliouras. "Spam filtering with naive Bayes." In *Proc. CEAS*, Jan. 2006, 28–69.
63. M. S. I. Mamun, M. A. Rathore, A. H. Lashkari, N. Stakhanova, and A. A. Ghorbani. "Detecting malicious URLs using lexical analysis." In *Network and System Security*, Cham, Switzerland: Springer, 2016, 467–482. https://doi.org/10.1007/978-3-319-46298-1_30.
64. M. Tavallaee, E. Bagheri, W. Lu, and A. A. Ghorbani. "A detailed analysis of the KDD CUP 99 data set." In *Proc. IEEE Symp. Comput. Intell. Secur. Defense Appl.*, Jul. 2009, 1–6. https://doi.org/ 10.1109/CISDA.2009.5356528.
65. A. Shiravi, H. Shiravi, M. Tavallaee, and A. A. Ghorbani. "Toward developing a systematic approach to generate benchmark datasets for intrusion detection." *Computer Security* 31, no. 3 (2012): 357–374. https://doi.org/10.1016/j.cose.2011.12.012.
66. C. Kolias, G. Kambourakis, A. Stavrou, and S. Gritzalis. "Intrusion detection in 802.11 networks: Empirical evaluation of threats and a public dataset." *IEEE Communications Surveys and Tutorials* 18, no. 1 (2016): 184–208, 1st. Quart., [Online]. https://doi.org/10.1109/COMST.2015.2402161
67. I. Sharafaldin, A. Habibi Lashkari, and A. A. Ghorbani. "Toward generating a new intrusion detection dataset and intrusion traffic characterization." In *Proc. 4th Int. Conf. Inf. Syst. Secur. Privacy*, 2018, 108–116. [Online]. https://www.researchgate.net/publication/322870768_Toward_Generating_a_New_Intrusion_Detection_Dataset_and_Intrusion_Traffic_Characterization
68. I. Sharafaldin, A. H. Lashkari, S. Hakak, and A. A. Ghorbani. "Developing realistic distributed denial of service (DDoS) attack dataset and taxonomy." In *Proc. Int. Carnahan Conf. Secur. Technol. (ICCST)*, Oct. 2019, 1–8. https://doi.org/10.1109/CCST.2019.8888419.
69. A. Alsaedi, N. Moustafa, Z. Tari, A. Mahmood, and A. Anwar. "TON_IoT telemetry dataset: A new generation dataset of IoT and IIoT for data-driven intrusion detection systems." *IEEE Access* 8 (2020): 165130–165150. https://doi.org/10.1109/ACCESS.2020.3022862.
70. R. Damasevicius, A. Venckauskas, S. Grigaliunas, J. Toldinas, N. Morkevicius, T. Aleliunas, and P. Smuikys. "LITNET-2020: An annotated real-world network flow dataset for network intrusion detection." *Electronics* 9, no. 5 (May, 2020): 800. https://doi.org/10.3390/electronics9050800.

71. G. Creech, and J. Hu. "Generation of a new IDS test dataset: Time to retire the KDD collection." In *Proc. IEEE Wireless Commun. Netw. Conf. (WCNC)*, Apr. 2013, 4487–4492. https://doi.org/10.1109/WCNC.2013.6555301.
72. N. Moustafa, and J. Slay. "UNSW-NB15: A comprehensive data set for network intrusion detection systems (UNSW-NB15 network data set)." In *Proc. Mil. Commun. Inf. Syst. Conf. (MilCIS)*, Nov. 2015, 1–6. https://doi.org/10.1109/MilCIS.2015.7348942.
73. S. García, M. Grill, J. Stiborek, and A. Zunino. "An empirical comparison of botnet detection methods." *Computer Security* 45 (Sep. 2014): 100–123. https://doi.org/10.1016/j.cose.2014.05.011.
74. S. Saad, I. Traore, A. Ghorbani, B. Sayed, D. Zhao, W. Lu, J. Felix, and P. Hakimian. "Detecting P2P botnets through network behavior."In *Proc. 9th Annu. Int. Conf. Privacy, Secur. Trust*, 2534 Jul. 2011, 174–180. https://doi.org/10.1109/PST.2011.5971980
75. N. Koroniotis, N. Moustafa, E. Sitnikova, and B. Turnbull. "Towards the development of realistic botnet dataset in the Internet of Things for network forensic analytics: Bot-IoT dataset." *Future Generation Computer Systems* 100 (Nov. 2019): 779–796. https://doi.org/10.1016/j.future.2019.05.041.
76. M. Zago, M. G. Pérez, and G. M. Pérez. "UMUDGA: A dataset for profiling DGA-based botnet." *Computer Security* 92 (May 2020): Art. no. 101719. https://doi.org/10.1016/j.cose.2020.101719.
77. R. Vinayakumar, K. P. Soman, P. Poornachandran, M. Alazab, and S. M. Thampi. "AmritaDGA: A comprehensive dataset for domain generation algorithms (DGAs) based domain name detection systems and application of deep learning." In *Big Data Recommender Systems*, vol. 2, O. Khalid, S. U. Khan, and A. Y. Zomaya, Eds. Stevenage, U.K.: Institution of Engineering and Technology, 2019, 455–485. https://doi.org/10.1049/PBPC035G_ch22.
78. D. Arp, M. Spreitzenbarth, M. Hübner, H. Gascon, and K. Rieck. "Drebin: Effective and explainable detection of Android malware in your pocket." NDSS, San Diego, CA, USA, *Tech. Rep.*, 2014. https://doi.org/10.14722/NDSS.2014.23247.
79. M. Kinkead, S. Millar, N. McLaughlin, and P. O'Kane. "Towards explainable CNNs for Android malware detection." *Procedia Computer Science* 184 (Jan. 2021): 959–965. https://doi.org/10.1016/j.procs.2021.03.118.
80. J. C. S. Reis, A. Correia, F. Murai, A. Veloso, and F. Benevenuto. "Explainable machine learning for fake news detection." In *Proc. 10th ACM Conf. Web Sci.*, New York, NY, USA, Jun. 2019, 17–26. https://doi.org/10.1145/3292522.3326027.
81. P. Hacker, R. Krestel, S. Grundmann, and F. Naumann. "Explainable AI under contract and tort law: Legal incentives and technical challenges." *Artificial Intelligence and Law* 28, no. 4 (Dec. 2020): 415–439. https://doi.org/10.1007/s10506-020-09260-6.
82. B. Mathew, P. Saha, S. M. Yimam, C. Biemann, P. Goyal, and A. Mukherjee. "HateXplain: A benchmark dataset for explainable hate speech detection." In *Proceedings of the AAAI Conference on Artificial Intelligence* 35, no. 17 (May 2021): 14867–14875.
83. H. Suryotrisongko, Y. Musashi, A. Tsuneda, and K. Sugitani. "Robust botnet DGA detection: Blending XAI and OSINT for cyber threat intelligence sharing." *IEEE Access* 10 (2022): 34613–34624. https://doi.org/10.1109/ACCESS.2022.3162588.
84. S. Araki, K. Takahashi, B. Hu, K. Kamiya, and M. Tanikawa. "Subspace clustering for interpretable botnet traffic analysis." In *Proc. IEEE Int. Conf. Commun. (ICC)*, May 2019, 1–6. https://doi.org/10.1109/ICC.2019.8761218.
85. M. Mazza, S. Cresci, M. Avvenuti, W. Quattrociocchi, and M. Tesconi. "RTbust: Exploiting temporal patterns for botnet detection on Twitter." In *Proc. 10th ACM Conf. Web Sci.*, New York, NY, USA, Jun. 2019, 183–192. https://doi.org/10.1145/3292522.3326015.

86. H. Bahsi, S. Nomm, and F. B. La Torre. "Dimensionality reduction for machine learning based IoT botnet detection." In *Proc. 15th Int. Conf. Control, Autom., Robot. Vis. (ICARCV)*, Nov. 2018, 1857–1862. https://doi.org/10.1109/ICARCV.2018.8581205.
87. P. P. Kundu, T. Truong-Huu, L. Chen, L. Zhou, and S. G. Teo. "Detection and classification of botnet traffic using deep learning with model explanation." *IEEE Trans. Dependable Secure Comput.*, early access, Jun. 15, 2022. https://doi.org/10.1109/TDSC.2022.3183361.
88. M. M. Alani. "BotStop: Packet-based efficient and explainable IoT botnet detection using machine learning." *Computer Communications* 193 (Sep. 2022): 53–62. https://doi.org/10.1016/j.comcom.2022.06.039.
89. Chai, Y., Y. Zhou, W. Li, and Y. Jiang. "An explainable multi-modal hierarchical attention model for developing phishing threat intelligence." *IEEE Transactions on Dependable and Secure Computing* 19, no. 2 (2022): 790–803. https://doi.org/10.1109/TDSC.2021.3119323.
90. Galego Hernandes, P. R., C. P. Floret, K. F. Cardozo De Almeida, V. C. Da Silva, J. P. Papa, and K. A. Pontara Da Costa. "Phishing detection using URL-based XAI techniques." In *Proceedings of the IEEE Symposium Series on Computational Intelligence (SSCI)*. 2021, 1–6. https://doi.org/10.1109/SSCI50451.2021.9659981.
91. Lin, Y., R. Liu, D. M. Divakaran, J. Y. Ng, Q. Z. Chan, Y. Lu, Y. Si, F. Zhang, and J. Dong. "Phishpedia: A hybrid deep learning-based approach to visually identify phishing webpages." In Proceedings of the 30th USENIX Security Symposium, 2021, 3793–3810. Retrieved July 21, 2022, from https://www.usenix.org/conference/usenixsecurity21/presentation/li.
92. Valecha, R., P. Mandaokar, and H. R. Rao. "Phishing email detection using persuasion cues." *IEEE Transactions on Dependable and Secure Computing* 19, no. 2 (2022): 747–756. https://doi.org/10.1109/TDSC.2021.3118931.
93. Barnard, P., N. Marchetti, and L. A. D. Silva. "Robust network intrusion detection through explainable artificial intelligence (XAI)." *IEEE Network Letters*, 2022. https://doi.org/10.1109/LNET.2022.3186589.
94. Andresini, G., A. Appice, F. P. Caforio, D. Malerba, and G. Vessio. "ROULETTE: A neural attention multi-output model for explainable network intrusion detection." *Expert Systems with Applications* 201 (2022): 117144. https://doi.org/10.1016/j.eswa.2022.117144.
95. Houda, Z. A. E., B. Brik, and L. Khoukhi. "'Why should I trust your IDS?': An explainable deep learning framework for intrusion detection systems in Internet of Things networks." *IEEE Open Journal of Communications Society* 3 (2022): 1164–1176. https://doi.org/10.1109/OJCOMS.2022.3188750.
96. Research Square. "Network intrusion detection based on explainable artificial intelligence." 2022. Retrieved July 21, 2022, from https://www.researchsquare.com.
97. Research Square. "KHO-XAI: Krill herd optimization and explainable artificial intelligence framework for network intrusion detection systems in Industry 4.0." June 10 2022. Retrieved July 21, 2022, from https://www.researchsquare.com.
98. Liu, H., C. Zhong, A. Alnusair, and S. R. Islam. "FAIXID: A framework for enhancing AI explainability of intrusion detection results using data cleaning techniques." *Journal of Network and Systems Management* 29, no. 4 (2021): 40. https://doi.org/10.1007/s10922-021-09606-8.
99. Mane, S., and D. Rao. "Explaining network intrusion detection system using explainable AI framework." 2021. *arXiv* preprint. https://arxiv.org/abs/2103.07110.
100. Zebin, T., S. Rezvy, and Y. Luo. "An explainable AI-based intrusion detection system for DNS over HTTPS (DoH) attacks." *IEEE Transactions on Information Forensics and Security* 17 (2022): 2339–2349. https://doi.org/10.1109/TIFS.2022.3183390.

101. Hsupeng, B., K.-W. Lee, T.-E. Wei, and S.-H. Wang. "Explainable malware detection using predefined network flow." *Proceedings of the 24th International Conference on Advanced Communication Technology (ICACT)*, 2022 February, 27–33. https://doi.org/10.23919/ICACT53585.2022.9728897.
102. Prasad, R. R., R. R. R. Robinson, C. Thomas, and N. Balakrishnan. "Evaluation of strategic decisions taken by autonomous agents using explainable AI." *Proceedings of the 4th International Conference on Security and Privacy (ISEA-ISAP)*, 2021, 1–8. https://doi.org/10.1109/ISEA-ISAP54304.2021.9689715.
103. Amarasinghe, K., K. Kenney, and M. Manic. "Toward explainable deep neural network-based anomaly detection." *Proceedings of the 11th International Conference on Human System Interaction (HSI)*, July 2018, 311–317. https://doi.org/10.1109/HSI.2018.8430788.
104. Dave, D., H. Naik, S. Singhal, and P. Patel. "Explainable AI meets healthcare: A study on heart disease dataset." 2020. *arXiv*:2011.03195.
105. Li, X., Y. Zhou, N. Dvornek, M. Zhang, S. Gao, J. Zhuang, D. Scheinost, L. H. Staib, P. Ventola, and J. S. Duncan. "BrainGNN: Interpretable brain graph neural network for fMRI analysis." *Medical Image Analysis* 74 (December 2021): 102233. https://doi.org/10.1016/j.media.2021.102233.
106. Hatwell, J., M. M. Gaber, and R. M. A. Azad. "Ada-WHIPS: Explaining AdaBoost classification with applications in the health sciences." *BMC Medical Informatics and Decision Making*, 20, no. 1 (December 2020): 1–25. https://bmcmedinformdecismak.biomedcentral.com/articles/10.1186/s12911-020-01201-2.
107. Chittajallu, D. R., B. Dong, P. Tunison, R. Collins, K. Wells, J. Fleshman, G. Sankaranarayanan, S. Schwaitzberg, L. Cavuoto, and A. Enquobahrie. "XAI-CBIR: Explainable AI system for content-based retrieval of video frames from minimally invasive surgery videos." *Proceedings of the IEEE 16th International Symposium on Biomedical Imaging (ISBI)*, April 2019, 66–69. https://doi.org/10.1109/ISBI.2019.8759428.
108. Stirnberg, R., J. Cermak, S. Kotthaus, M. Haeffelin, H. Andersen, J. Fuchs, M. Kim, J.-E. Petit, and O. Favez. "Meteorology-driven variability of air pollution (PM1) revealed with explainable machine learning." *Atmospheric Chemistry and Physics* 21, no. 5 (March 2021): 3919–3948. https://doi.org/10.5194/acp-21-3919-2021.
109. Monje, L., R. A. Carrasco, C. Rosado, and M. Sánchez-Montañés. "Deep learning XAI for bus passenger forecasting: A use case in Spain." *Mathematics* 10, no. 9 (April 2022): 1428. https://doi.org/10.3390/math10091428.
110. Kostopoulos, G., T. Panagiotakopoulos, S. Kotsiantis, C. Pierrakeas, and A. Kameas. "Interpretable models for early prediction of certification in MOOCs: A case study on a MOOC for smart city professionals." *IEEE Access* 9 (2021): 165881–165891. https://doi.org/10.1109/ACCESS.2021.3134787.
111. Feng, Y., D. Wang, Y. Yin, Z. Li, and Z. Hu. "An XGBoost-based casualty prediction method for terrorist attacks." *Complex & Intelligent Systems* 6, no. 3 (October 2020): 721–740. https://doi.org/10.1007/s40747-020-00173-0.
112. Garrido, M. C., J. M. Cadenas, A. Bueno-Crespo, R. Martínez-España, J. G. Giménez, and J. M. Cecilia. "Evaporation forecasting through interpretable data analysis techniques." *Electronics* 11, no. 4 (February 2022): 536. https://doi.org/10.3390/electronics11040536.
113. Viana, C. M., M. Santos, D. Freire, P. Abrantes, and J. Rocha. "Evaluation of the factors explaining the use of agricultural land: A machine learning and model-agnostic approach." *Ecological Indicators* 131 (November 2021): 108200. https://doi.org/10.1016/j.ecolind.2021.108200.

114. Daníelsson, J., R. Macrae, and A. Uthemann. "Artificial intelligence and systemic risk." *Journal of Banking & Finance* 140 (July 2022): 106290. https://doi.org/10.1016/j.jbankfin.2021.106290.
115. Kute, D. V., B. Pradhan, N. Shukla, and A. Alamri. "Deep learning and explainable artificial intelligence techniques applied for detecting money laundering—A critical review." *IEEE Access* 9 (2021): 82300–82317. https://doi.org/10.1109/ACCESS.2021.3086230.
116. Sachan, S., J.-B. Yang, D.-L. Xu, D. E. Benavides, and Y. Li. "An explainable AI decision-support-system to automate loan underwriting." *Expert Systems with Applications* 144 (April 2020): 113100. https://doi.org/10.1016/j.eswa.2019.113100.
117. Yang, L., E. M. Kenny, T. L. J. Ng, Y. Yang, B. Smyth, and R. Dong. "Generating plausible counterfactual explanations for deep transformers in financial text classification." 2020. *arXiv*:2010.12512. Retrieved from https://arxiv.org/abs/2010.12512.
118. Hanif, A. "Towards explainable artificial intelligence in banking and financial services." 2021. *arXiv*:2112.08441. Retrieved from https://arxiv.org/abs/2112.08441
119. Gurcan, F., N. E. Cagiltay, and K. Cagiltay. "Mapping human–computer interaction research themes and trends from its existence to today: A topic modeling-based review of past 60 years." *International Journal of Human-Computer Interaction* 37, no. 3 (February 2021): 267–280. https://doi.org/10.1080/10447318.2020.1819668.
120. Shams Amiri, S., S. Mottahedi, E. R. Lee, and S. Hoque. "Peeking inside the black-box: Explainable machine learning applied to household transportation energy consumption." *Computers, Environment and Urban Systems* 88 (2021): 101647. https://doi.org/10.1016/j.compenvurbsys.2021.101647.
121. Bustos, C., D. Rhoads, A. Solé-Ribalta, D. Masip, A. Arenas, A. Lapedriza, and J. Borge-Holthoefer. "Explainable, automated urban interventions to improve pedestrian and vehicle safety." *Transportation Research Part C: Emerging Technologies* 125 (April 2021): 103018. https://doi.org/10.1016/j.trc.2021.103018.

Index

T

U

V

W

X

For Product Safety Concerns and Information please contact our
EU representative GPSR@taylorandfrancis.com Taylor & Francis
Verlag GmbH, Kaufingerstraße 24, 80331 München, Germany